Descriptive Geometry Worksheets
with Computer Graphics

Series A

9th Edition

E. G. Paré
Lecturer at Loyola Marymount College

R. O. Loving
Professor Emeritus of Engineering Graphics and Formerly Chairman of the Department Illinois Institute of Technology

I. L. Hill
Professor Emeritus of Engineering Graphics and Formerly Chairman of the Department Illinois Institute of Technology

R. C. Paré
Associate Professor of Mechanical Engineering Technology University of Houston, Houston, Texas

PRENTICE HALL, Upper Saddle River, New Jersey 07458

Acquisitions editor: *Eric Svendsen*
Production editor: *Barbara Kraemer*
Production supervisor: *Barbara Murray*
Production Coordinator: *Julia Meehan*

 ©1997 by Prentice-Hall, Inc.
Upper Saddle River, New Jersey 07458

All rights reserved. No part of this book may be reproduced, in any form or by any means, without permission in writing from the publisher.

Printed in the United States of America

10 9 8 7 6 5 4 3

ISBN 0-02-391342-8

Prentice-Hall International (UK) Limited, *London*
Prentice-Hall of Australia Pty. Limited, *Sydney*
Prentice-Hall Canada Inc., *Toronto*
Prentice-Hall Hispanoamericana, S.A., *Mexico*
Prentice-Hall of India Private Limited, *New Delhi*
Prentice-Hall of Japan, Inc., *Tokyo*
Prentice-Hall Asia Pte. Ltd., *Singapore*
Editora Prentice-Hall do Brasil, Ltda., *Rio de Janeiro*

Symbols for Use on Drawings

TL — True Length
EV — Edge View
TS — True Size
LI — Line of Intersection

— Parallel

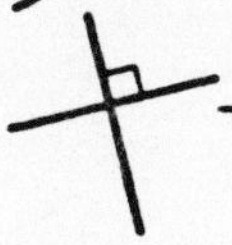

— Perpendicular

— Piercing Point of line and surface

Symbols for Instructor's Corrections

C — Show construction
D — Show dimensions; show given or required data
I — Improve form or spacing
H — Too heavy
NH — Not heavy enough
ND — Not dark enough
SL — Sharpen pencil or compass lead
GL — Use guide lines
A — Improve arrowheads

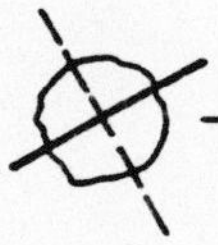

— Error in encircled area

Equivalents

inches to millimeters		millimeters to inches	
in.	mm	mm	in.
0.001	0.025	1	0.039
0.002	0.051	2	0.079
0.003	0.076	3	0.118
0.004	0.102	4	0.158
0.005	0.127	5	0.197
0.006	0.152	6	0.236
0.007	0.178	7	0.276
0.008	0.203	8	0.315
0.009	0.229	10	0.394
0.010	0.254	12	0.472
0.020	0.508	16	0.630
0.030	0.762	20	0.787
0.040	1.016	25	0.984
0.050	1.270	30	1.181
0.060	1.524	35	1.378
0.070	1.778	40	1.575
0.080	2.032	45	1.772
0.090	2.286	50	1.968
0.100	2.540	55	2.165
0.200	5.080	60	2.362
0.300	7.620	65	2.559
0.400	10.160	70	2.756
0.500	12.700	75	2.953
0.600	15.240	80	3.150
0.700	17.780	85	3.346
0.800	20.320	90	3.543
0.900	22.860	95	3.740
1.000	25.400	100	3.937

LENGTH	1 millimeter (mm) = .0393701 inch (in.) 1 inch (in.) = 25.4 millimeter (mm)
AREA	1 square millimeter (mm^2) = 0.00155 square inch (in.^{2}) 1 square inch (in.^{2}) = 645.16 square millimeter (mm^2)
MASS	1 kilogram (kg) = 2.20462 pound (lb) 1 pound (lb) = 0.453592 kilogram (kg)
FORCE	1 Newton (N) = 9.80665 kg·m/s^2

Preface

Descriptive Geometry Worksheets, Series A is intended primarily for use with Descriptive Geometry, 9th Edition by the same authors and also published by Prentice-Hall. The group number identifications of the individual worksheets parallel the chapters of that textbook. However, this workbook may be used with any good descriptive geometry or comprehensive engineering graphics text.

Time available for teaching is limited. The objective of this workbook is to eliminate repetitious drawing and setup and allow the student to spend time learning theory and application. It is intended that the problems be solved by the widely accepted "direct" method. The layouts are designed so that either "folding line" or "reference plane" notation may be used. Good drafting practices should be followed to produce acceptable accuracy. All construction lines should be made lightly and should not be erased.

This workbook of descriptive geometry problems has been prepared by the authors, as an answer to the question frequently posed by students: "What use will I make of descriptive geometry?" Many practical applications presented in the problems speak for themselves. With but a few exceptions each worksheet contains at least one interesting practical application, many of which were suggested by a variety of industrial concerns.

Practical problems have not been used merely because they are practical, however. They are used only where the authors consider that they teach the theory involved at least as well as the corresponding abstract exercise. In general, each new division of subject matter is introduced with theoretical problems to focus the student's attention on the fundamentals involved. These are usually followed by applications to practical situations to make obvious the utility of the subject matter.

The use of many pictorial illustrations and the problems in pictorial form serve to stimulate interest and to contribute toward the development of three-dimensional visualization.

As in the case with the new 9th edition of the textbook, additional worksheets have been included in Group 14 to reflect new material in vectors, beams and trusses.

A number of problems are given in Group 24 to provide the student with essential training in problem layout. These problems also serve as comprehensive review material, since their solutions require the use of a number of the fundamentals.

A selected sample of computer graphics projects has been provided in Group 25. The computer graphics solutions demonstrate how mathematics and the computer can be combined to provide solutions to problems solved manually in the earlier groups.

To conform to the latest trends, all measurements are given with metric values. A conversion table and some metric scales are included for the student's convenience.

Since Series B of the Descriptive Geometry Worksheets also follows the same organization pattern, either workbook may be used with only minor schedule changes.

E.G.P.
R.O.L.
I.L.H.
R.C.P.

Contents

Text references may be found in Paré, Loving, Hill, and Paré, *Descriptive Geometry*, Ninth Edition (Prentice Hall, 1997). For convenient reference, the chapter numbers and titles in the text correspond to the group numbers and titles herein.

Notation (inside front cover)
Metric Scales—Millimeters (inside back cover)

A Lowercase Letters and Notation

GROUP 1
Orthographic Projection
1a   Visibility
1b   Visibility

GROUP 2
Primary Auxiliary Views
2a   Auxiliary Views
2b   Auxiliary Views

GROUP 3
Lines
3a   True Lengths and True Angles
3b   True Length Application
3c   Intersecting Lines

GROUP 4
Planes
4a   Lines on Planes
4b   Lines on Planes

GROUP 5
Successive Auxiliary Views
5a   Point View of a Line
5b   Normal View of a Plane

GROUP 6
Piercing Points
6a   Piercing Points
6b   Piercing Points
6c   Pictorial Intersections
6d   Pictorial Intersections

GROUP 7
Intersection of Planes
7a   Intersection of Planes
7b   Pictorial Intersections
7c   Pictorial Intersections

GROUP 8
Angle Between Planes
8a   Dihedral Angles

GROUP 9
Parallelism
9a   Parallelism
9b   Pictorial Parallelism

GROUP 10
Perpendicularity
10a   Perpendicularity
10b   Perpendicularity
10c   Common Perpendicular

GROUP 11
Angle Between Line and Oblique Plane
11a   Angle Between a Line and Plane

GROUP 12
Mining and Civil Engineering Problems
12a   Mining and Geology
12b   Contour Map and Outcrop
12c   Cut and Fill

GROUP 13
Revolution
13a   Revolution of Lines

GROUP 14
Concurrent Vectors
14a Vectors
14b Non Coplanar Vectors
14c Non Concurrent Vectors
14d Truss Stresses

GROUP 15
Plane Tangencies
15a Tangent Planes
15b Pictorial Tangencies

GROUP 16
Intersections of Planes and Solids
16a Intersection of a Plane and a Solid
16b Intersection of a Plane and a Solid

GROUP 17
Developments
17a Radial Line Development
17b Transition Unit

GROUP 18
Intersections of Surfaces
18a Polyhedral Intersections
18b Intersection of Forms
18c Pictorial Intersection

GROUP 19
Shades and Shadows
19a Shades and Shadows
19b Shade and Shadow
19c Pictorial Shades and Shadows

GROUP 20
Pictorial Projection
20a Perspective Pictorial
20b Perspective
20c Perspective Intersections
20d Perspective Shade and Shadow

GROUP 21
Conics
21a Conics

GROUP 22
Map Projections
22a Map Projections

GROUP 23
Spherical Triangles
23a Spherical Triangle
23b Navigation Project

GROUP 24
Review
24a Basic Review Projects
24b Basic Review Projects
24c Special Intersections
24d Shade, Shadow, & Intersection
24e Intersection, Shade, & Shadow
24f Golf Course Layout
24g Skateboard Facility
24h Special Intersections
24i Non-Coplanar Vectors

GROUP 25
Computer Graphics
25a True Length, Bearing, and Grade of a Line
25b Line Intersecting a Plane
25c Pictorial Line Intersecting a Plane
25d Pictorial Intersecting Planes
25e Intersecting Planes
25f Development
25g Isometric Shade and Shadow
25h Perspective Shade and Shadow

Blank Layout Sheets

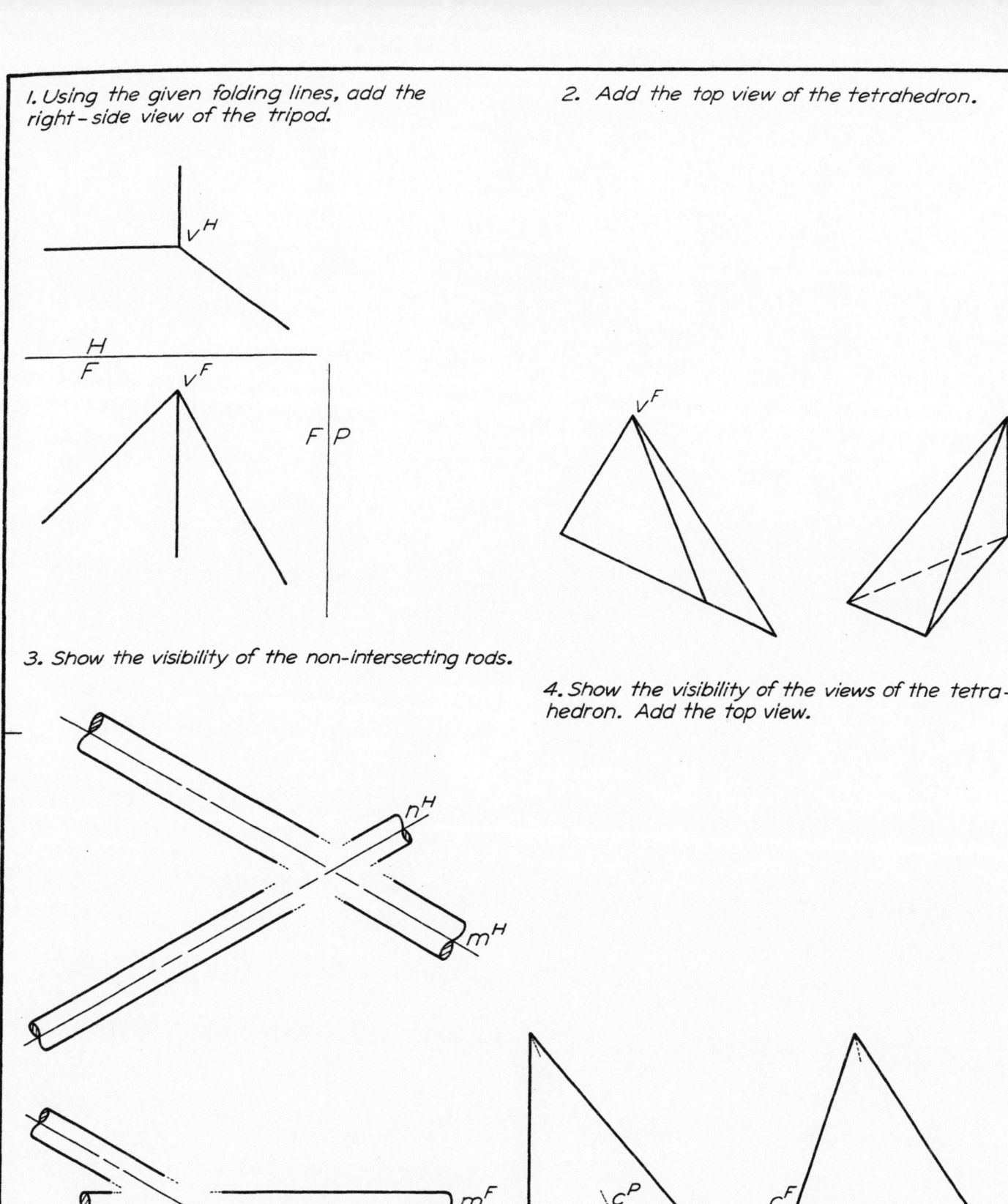

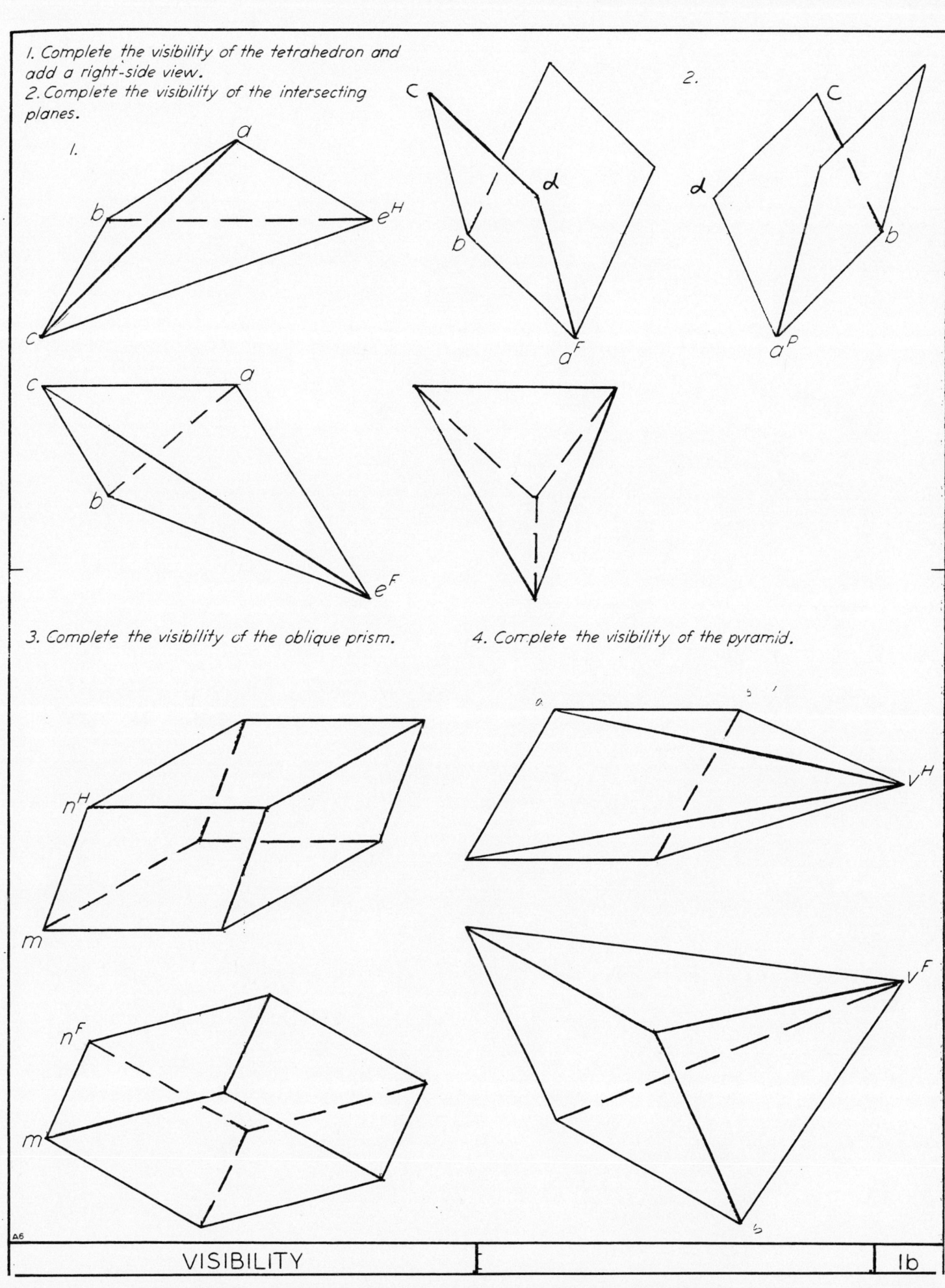

1. Draw complete views P and I. (Identical transfer distances are employed in the construction of views P and I.) Draw view 2.

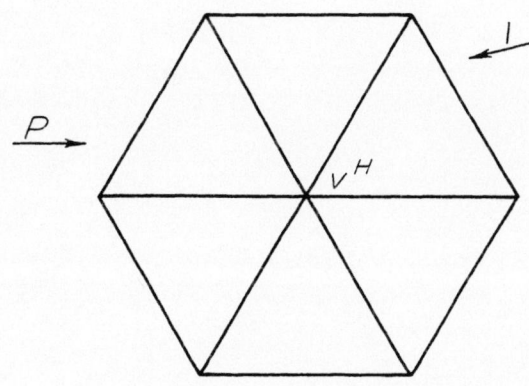

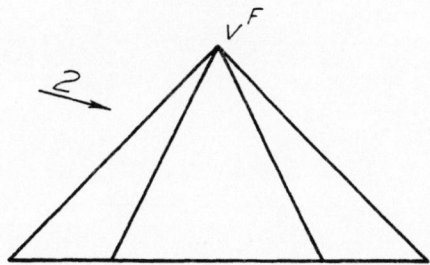

2. Add view I.

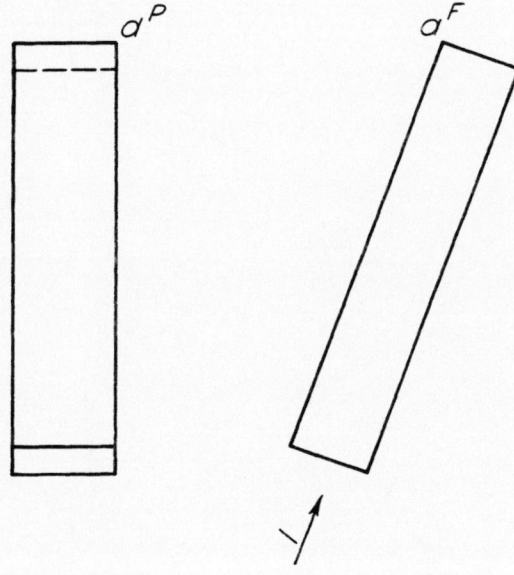

3. Add views P and I of the triangular plane.

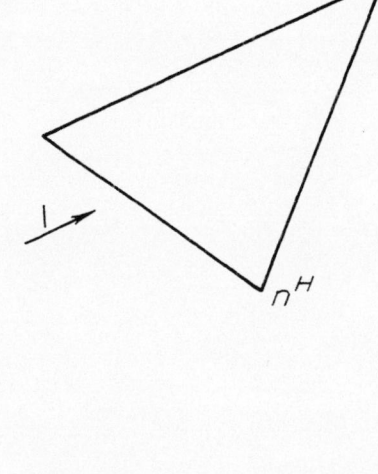

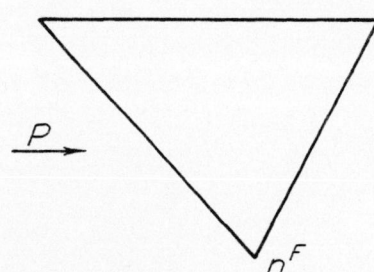

PRIMARY AUXILIARY VIEWS — 2a

1. Show the true size of triangle ABC and calculate the area of a circle that will circumscribe the triangle.
Scale = 1:400   AREA: _____ m²

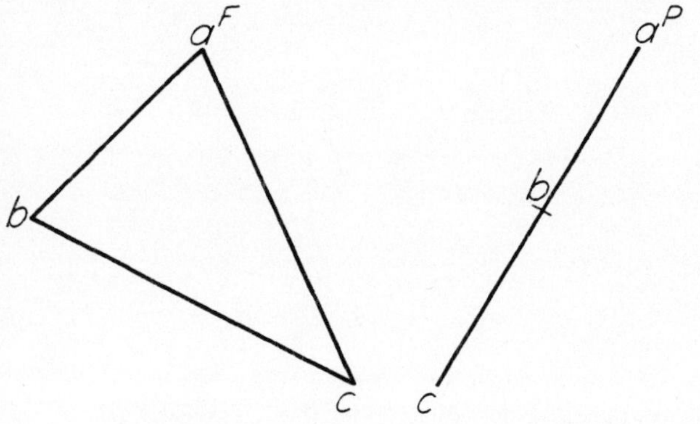

2. Show the true size of parallelogram MNOS and calculate its area. Scale = 1:80
AREA: _____ m²

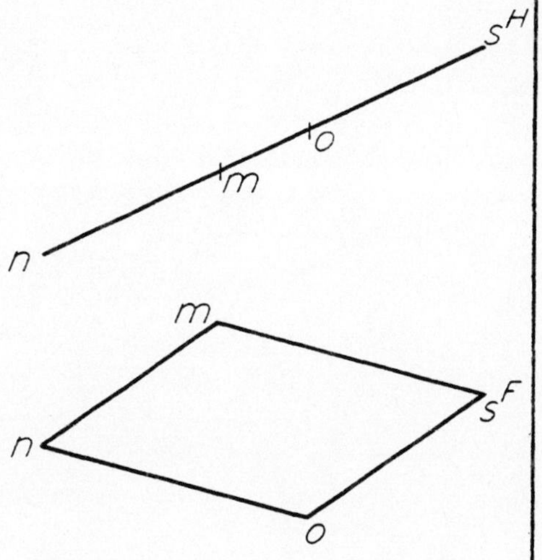

3. Show the true size of plane VAE and plane ABKV. Scale and compute these two areas. Scale = 1:500
AREA - VAE: _____ m² ; ABKV: _____ m²

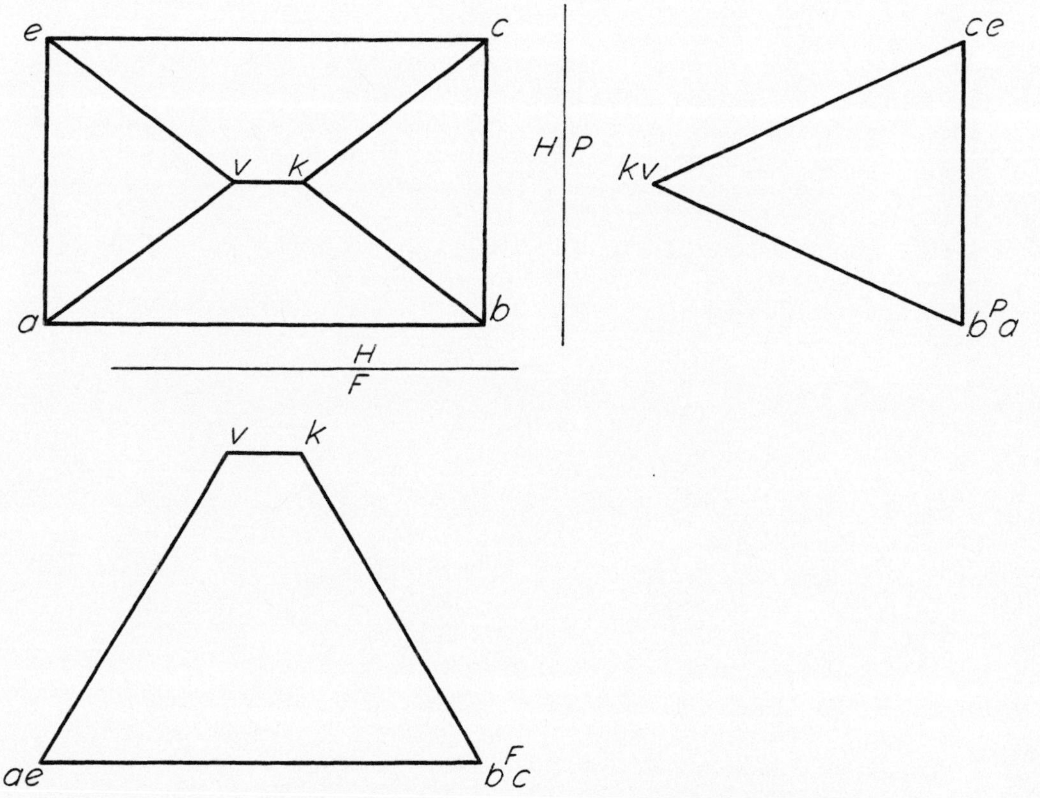

NORMAL VIEW OF PLANE

2b

1. Show and identify the true length (TL) of each of the tripod legs. Scale = 1:40

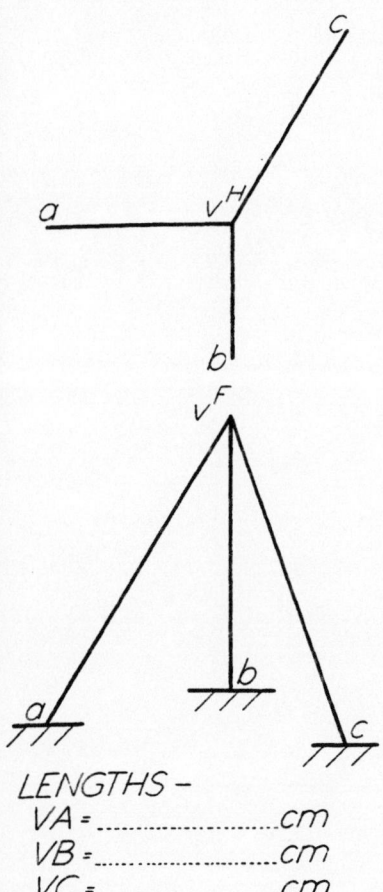

LENGTHS –
VA = ...cm
VB = ...cm
VC = ...cm

2. Establish the true lengths of the edges MN, NK, and VK. Scale = 1:200

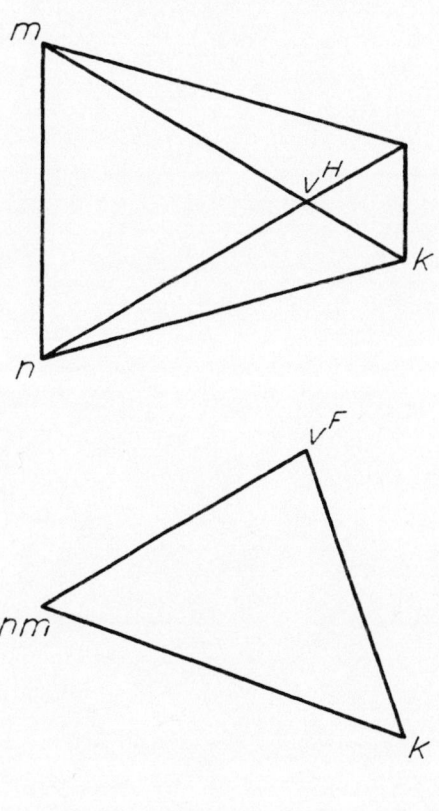

LENGTHS –
MN = ...m
NK = ...m
VK = ...m

3. For member VE obtain the true angle (θ_F) with the frontal wall. For member VO obtain the true angle (θ_H) with the horizontal surface.

ANGLES –
VE (θ_F) = ...
VO (θ_H) = ...

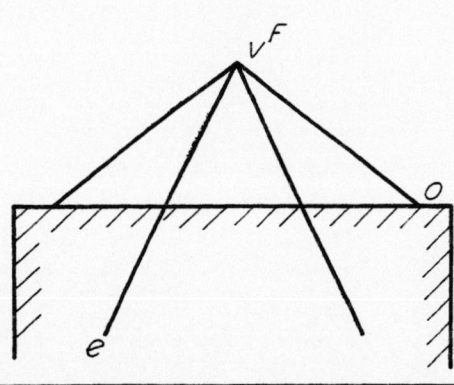

HORIZONTAL SURFACE
FRONTAL WALL

TRUE LENGTHS & ANGLES — 3a

1. Indicate the bearing and determine the length and grade of a ski-tow cable MN.
Scale 1:10 000

2. Pipeline AB bears N120° on a downgrade of 32° for a distance of 3500 meters. Complete the views.
Scale 1:50 000

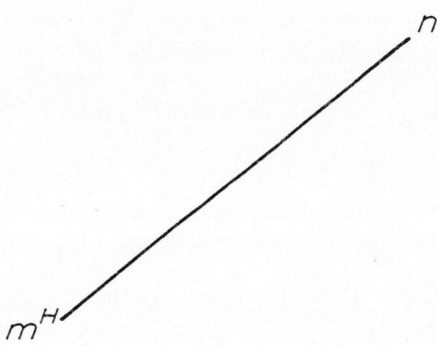

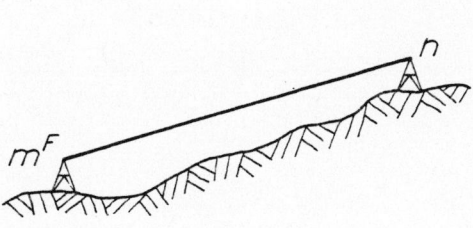

Bearing = ...
Length = ...
Grade   = ...

3. Determine the angles which the members of the SUPPORT FRAME make with the surfaces to which they are attached. Find the true length of each member.   Scale 1:100

VA: TL = ...
    θ = ...
VB: TL = ...
    θ = ...

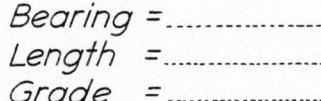

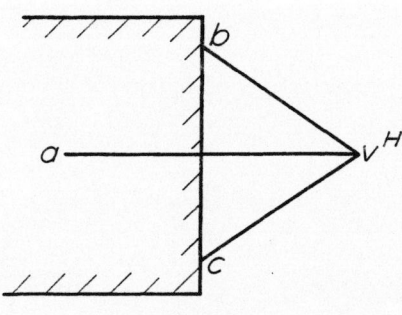

TRUE LENGTH APPLICATIONS — 3b

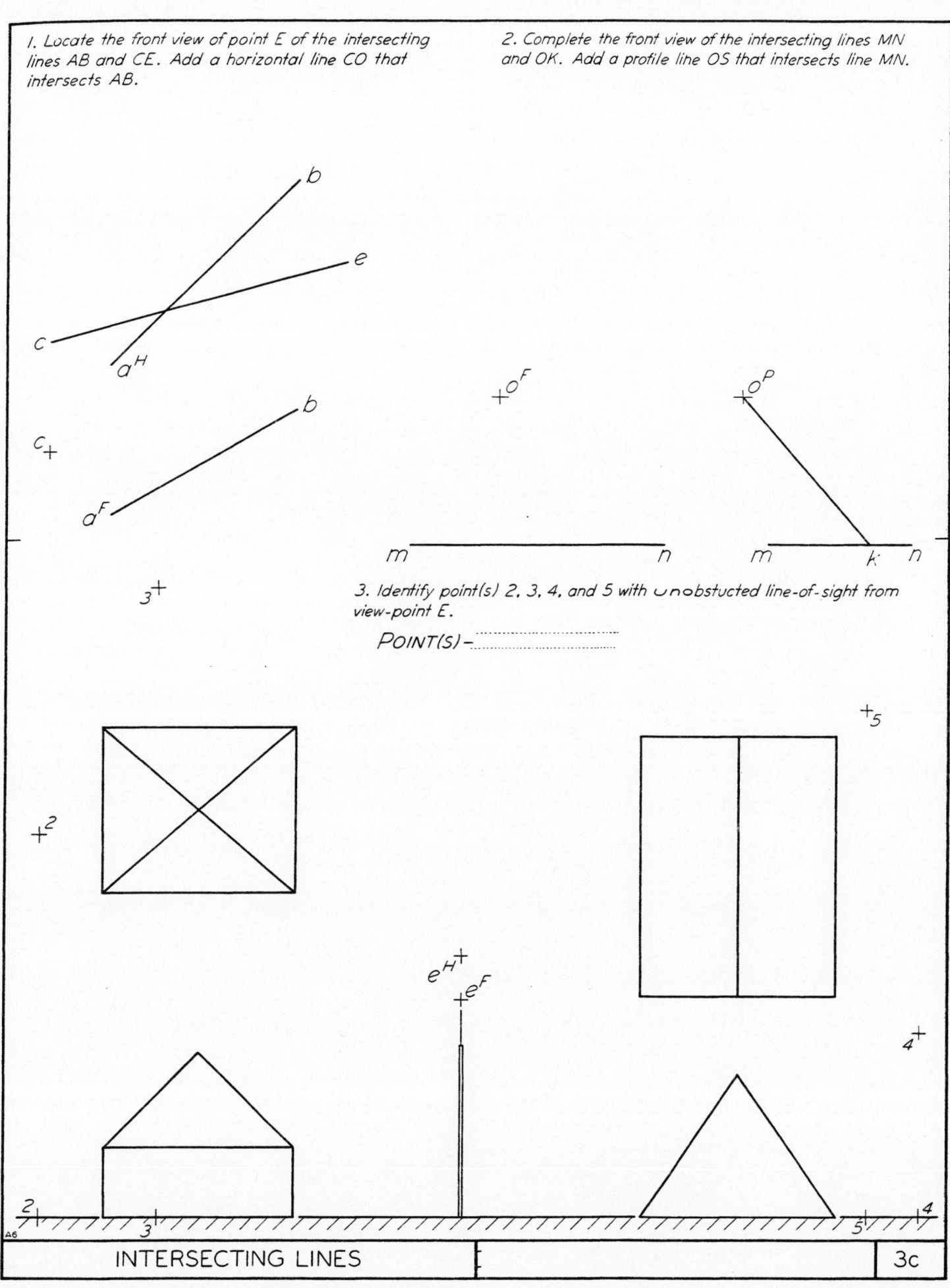

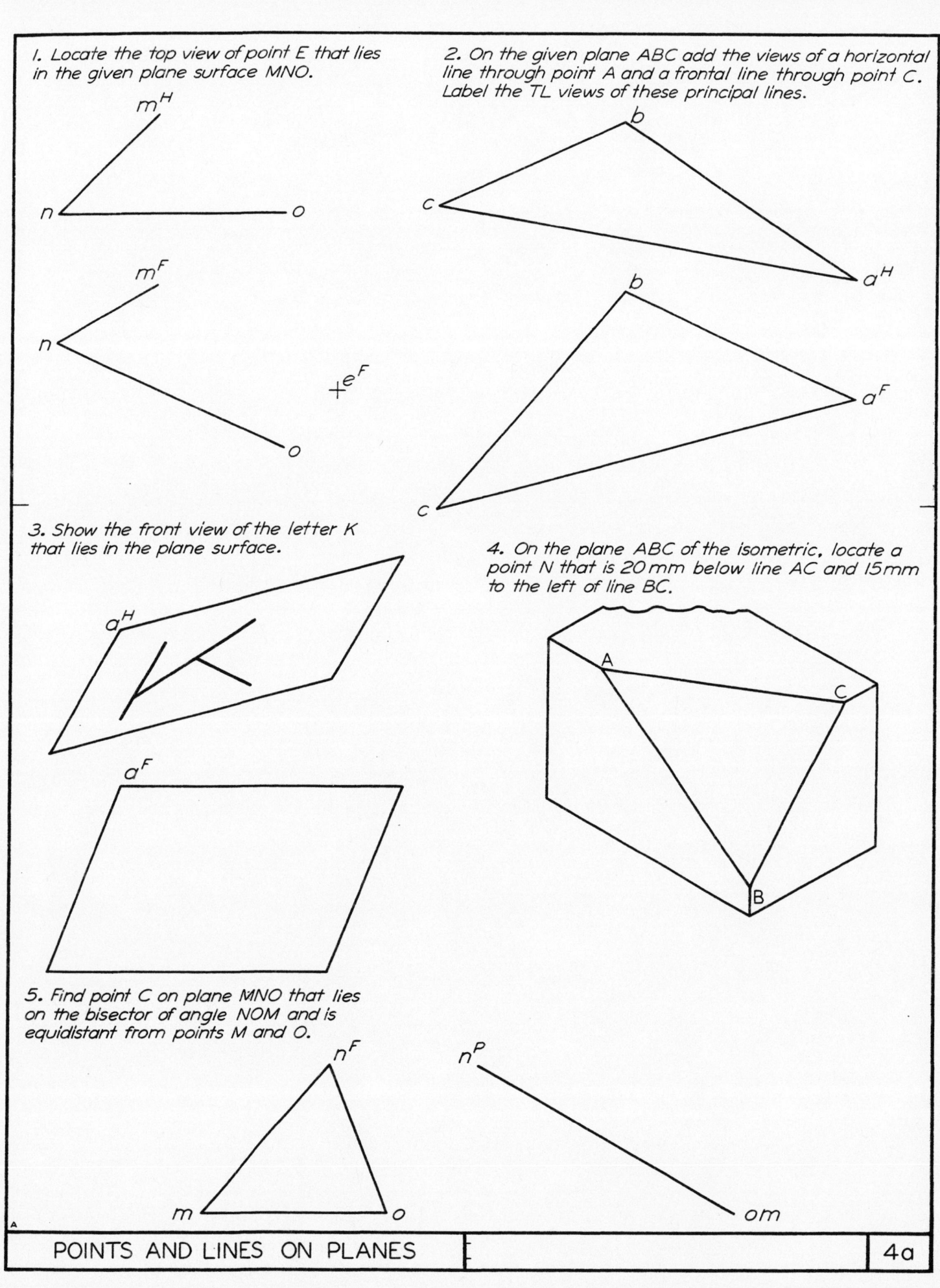

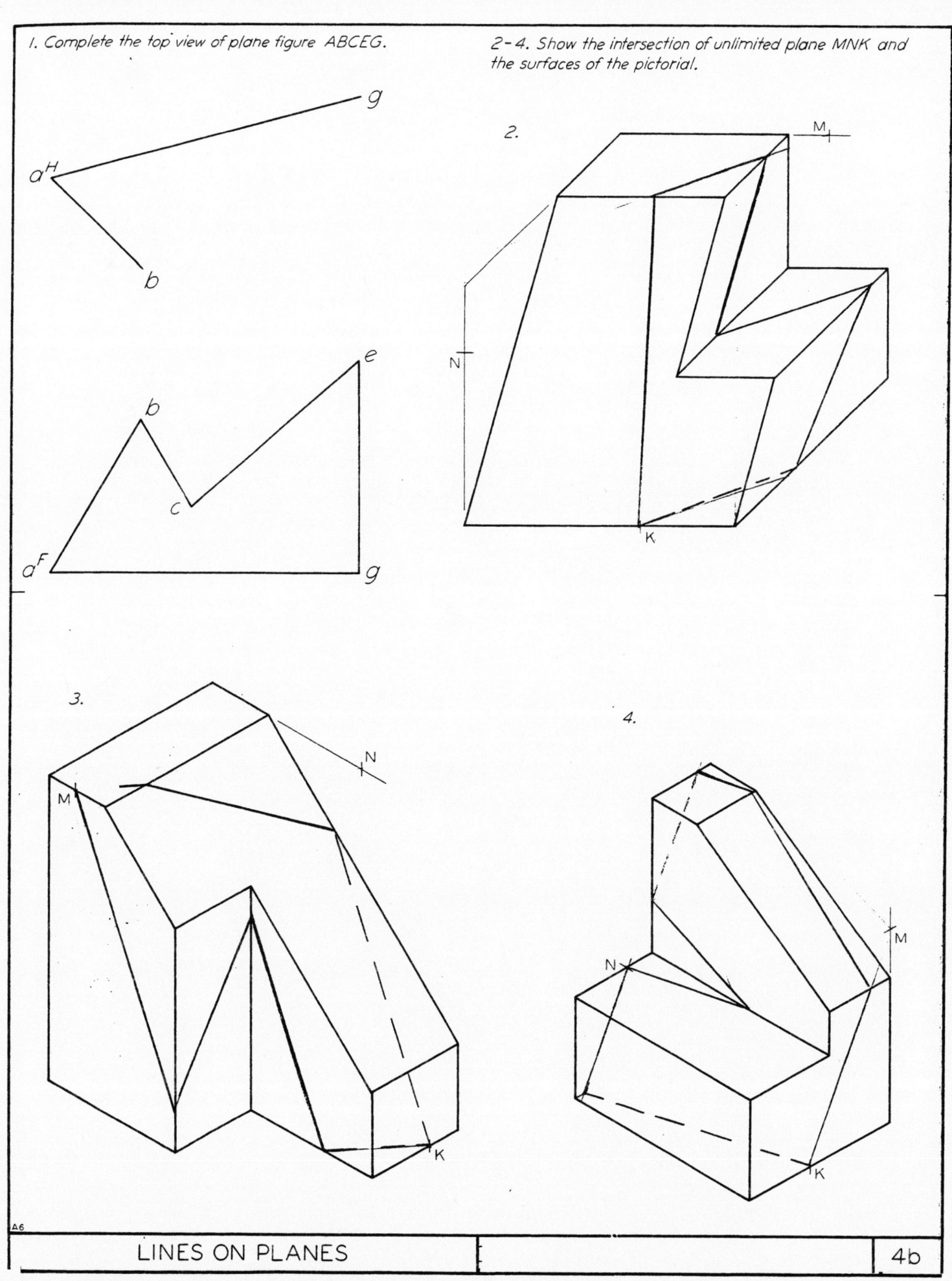

1. Show a point view of line AB.

2. Establish a point view of line CE.

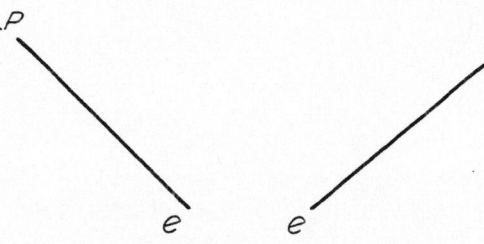

3. Obtain the distance from the surface of the sphere centered at point O to the line MN. Scale = 1:40

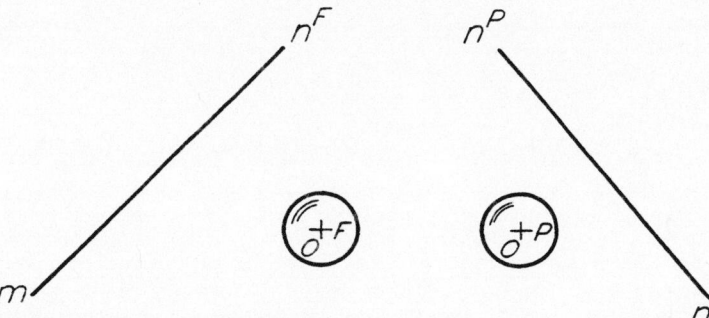

DISTANCE - _____ cm

4. Show a view of the prism that will include a point view of line EG.

5. Show a view of the tetrahedron that includes a point view of line VA. Show a view that includes a point view of line VC.

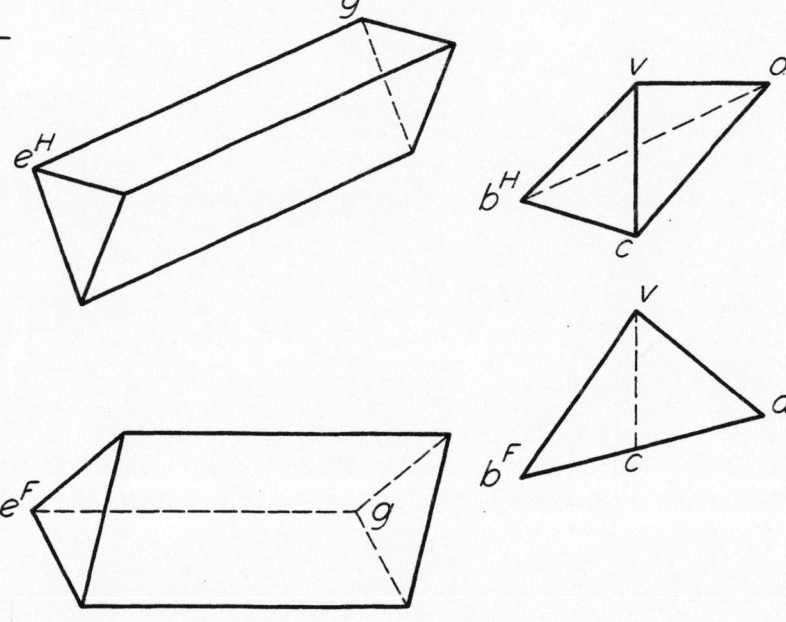

POINT VIEW OF LINE   5a

1. Show the true size of the triangle ABC and compute the area.

   AREA = ... mm^2

2. Show the true size of planes MNO and VNO of the tetrahedron.

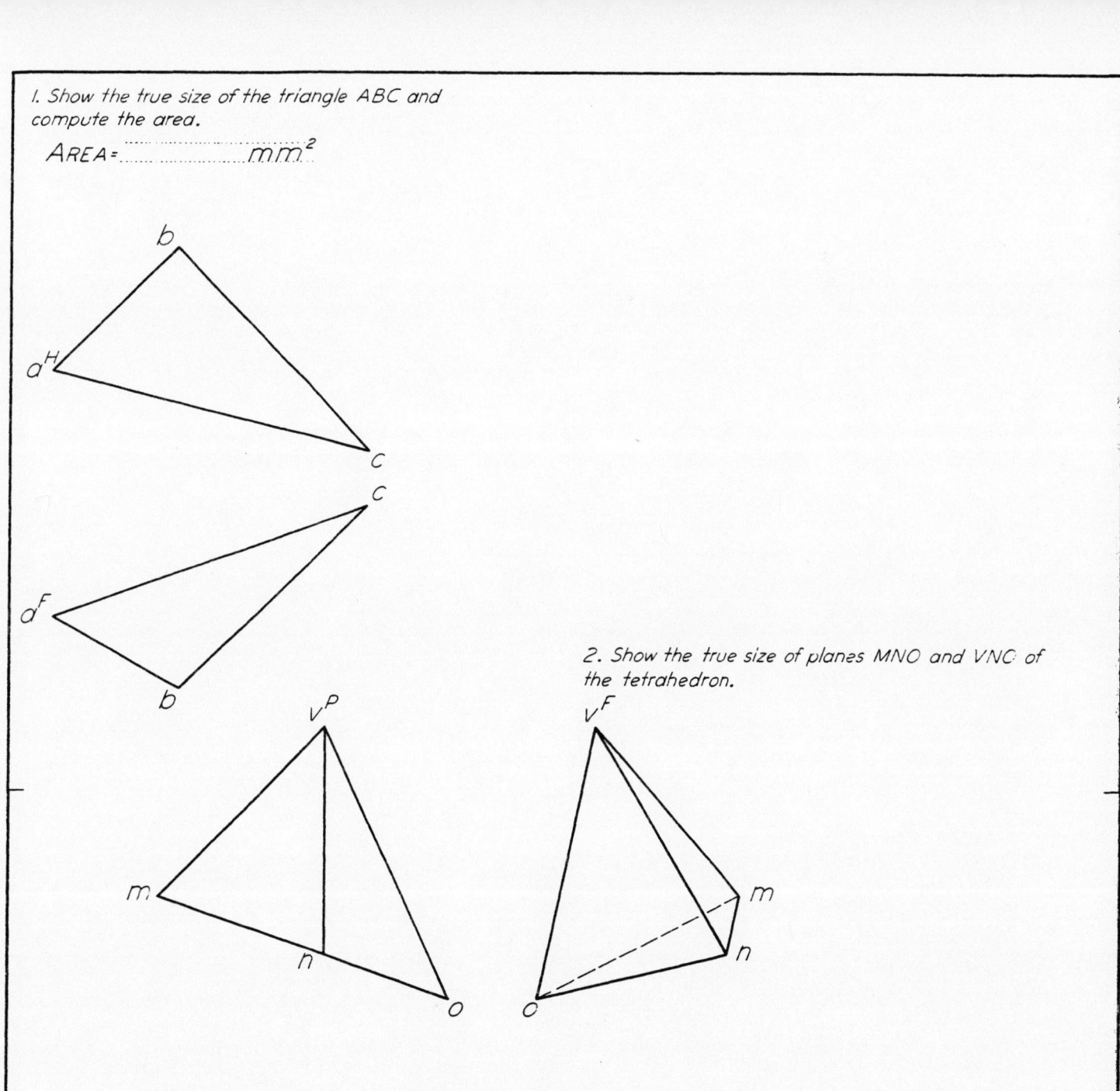

NORMAL VIEW OF PLANE — 5b

NOTE: On this sheet use the auxiliary-view method and show complete visibility.

1. Show the intersection of the line AB and the plane.

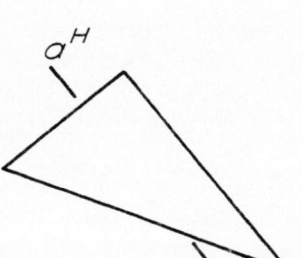
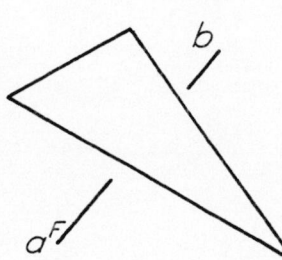

2. Obtain the intersection of line MN and the plane surface.

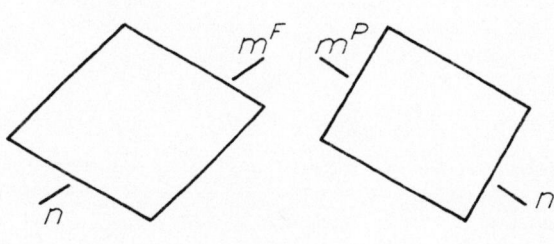

3. Find the intersection of CE and the surfaces of the pyramid. Omit that portion of CE within the solid.

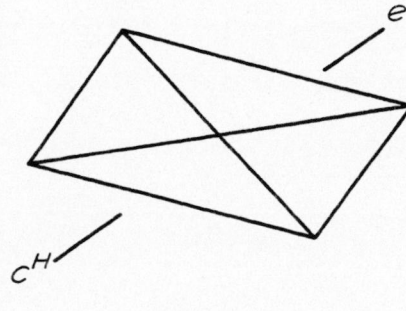
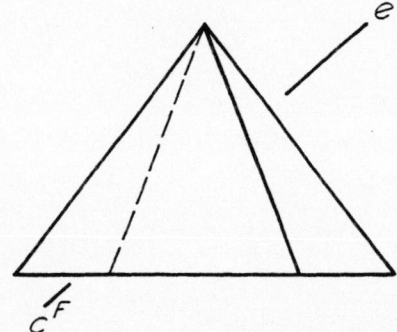

4. Complete the front view of line MO so that it will pierce plane VAB and base ABCE.

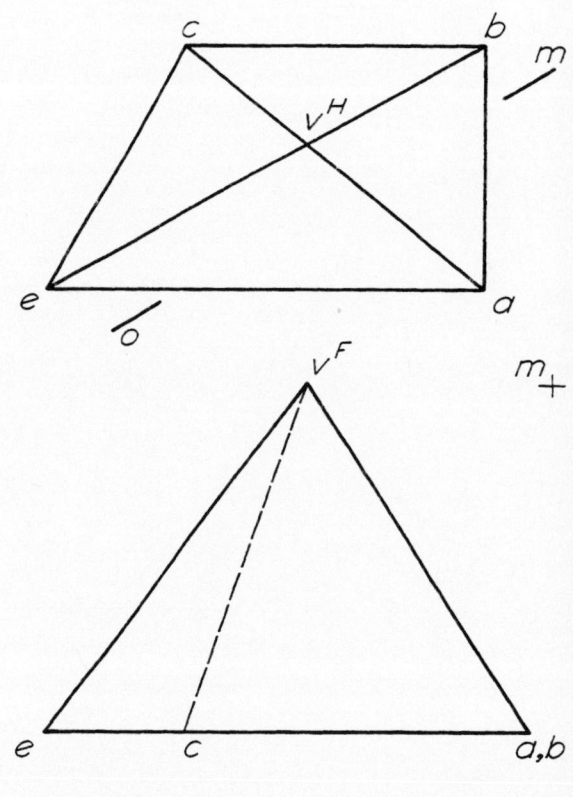

PIERCING POINTS 6a

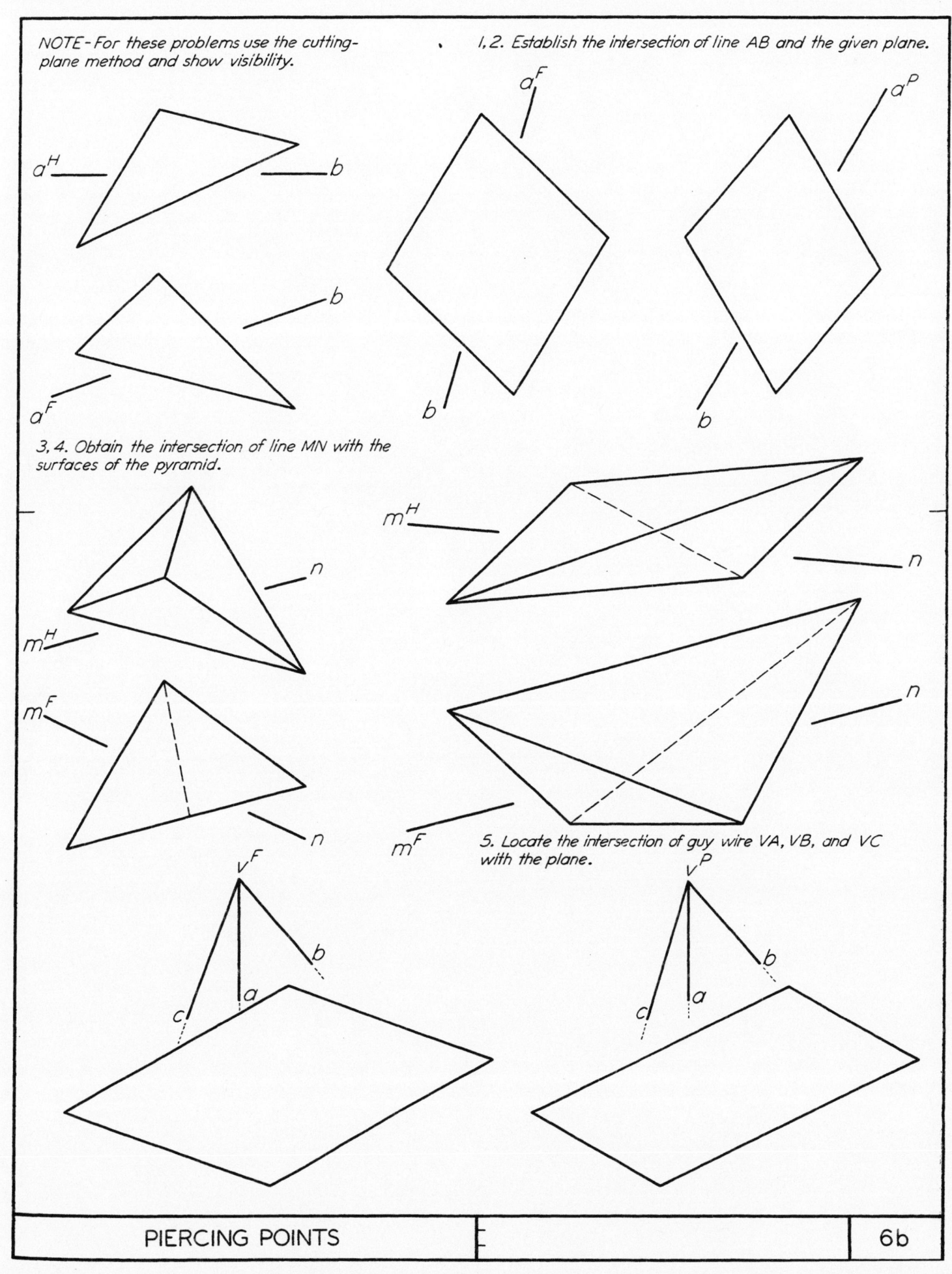

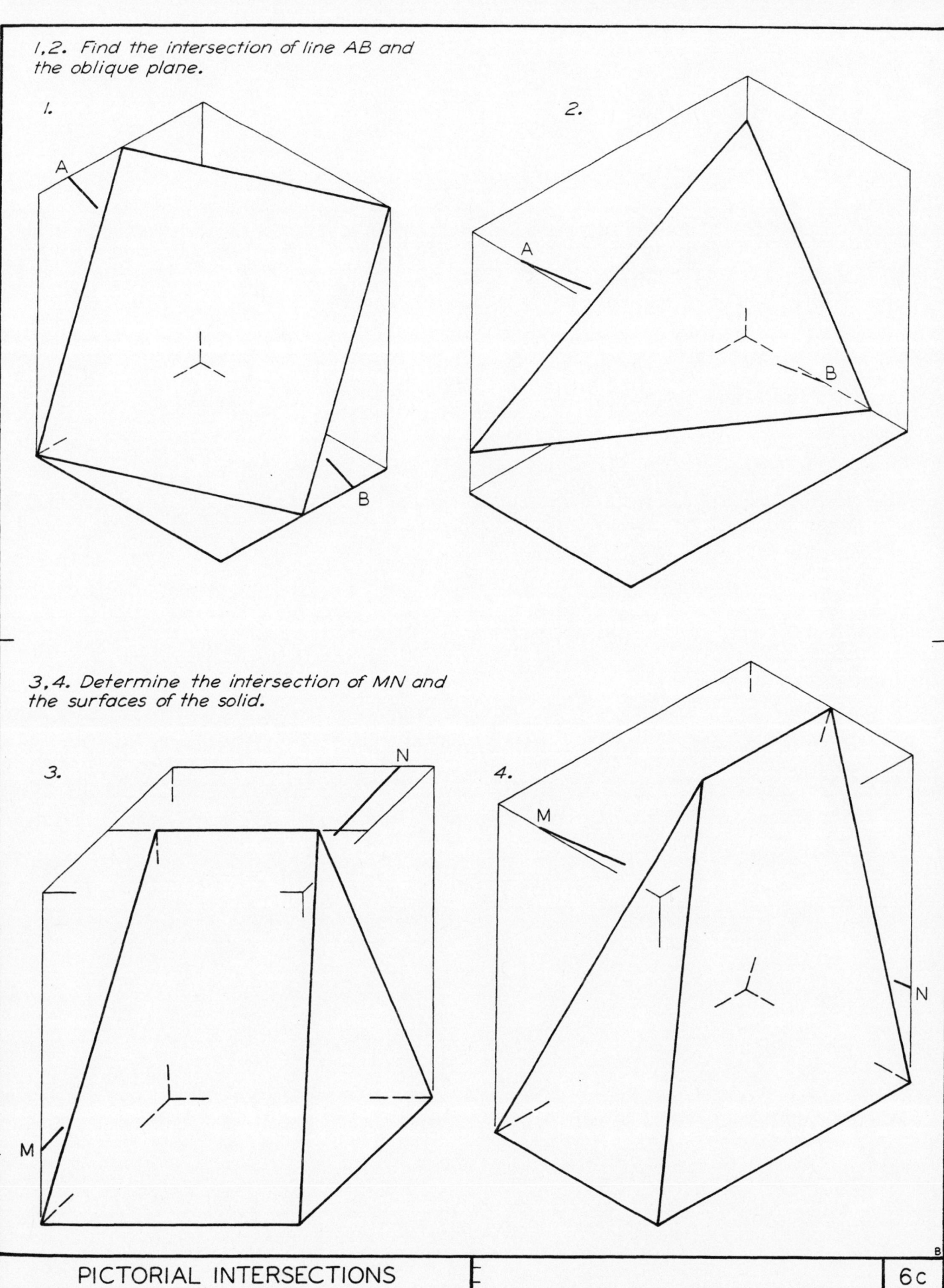

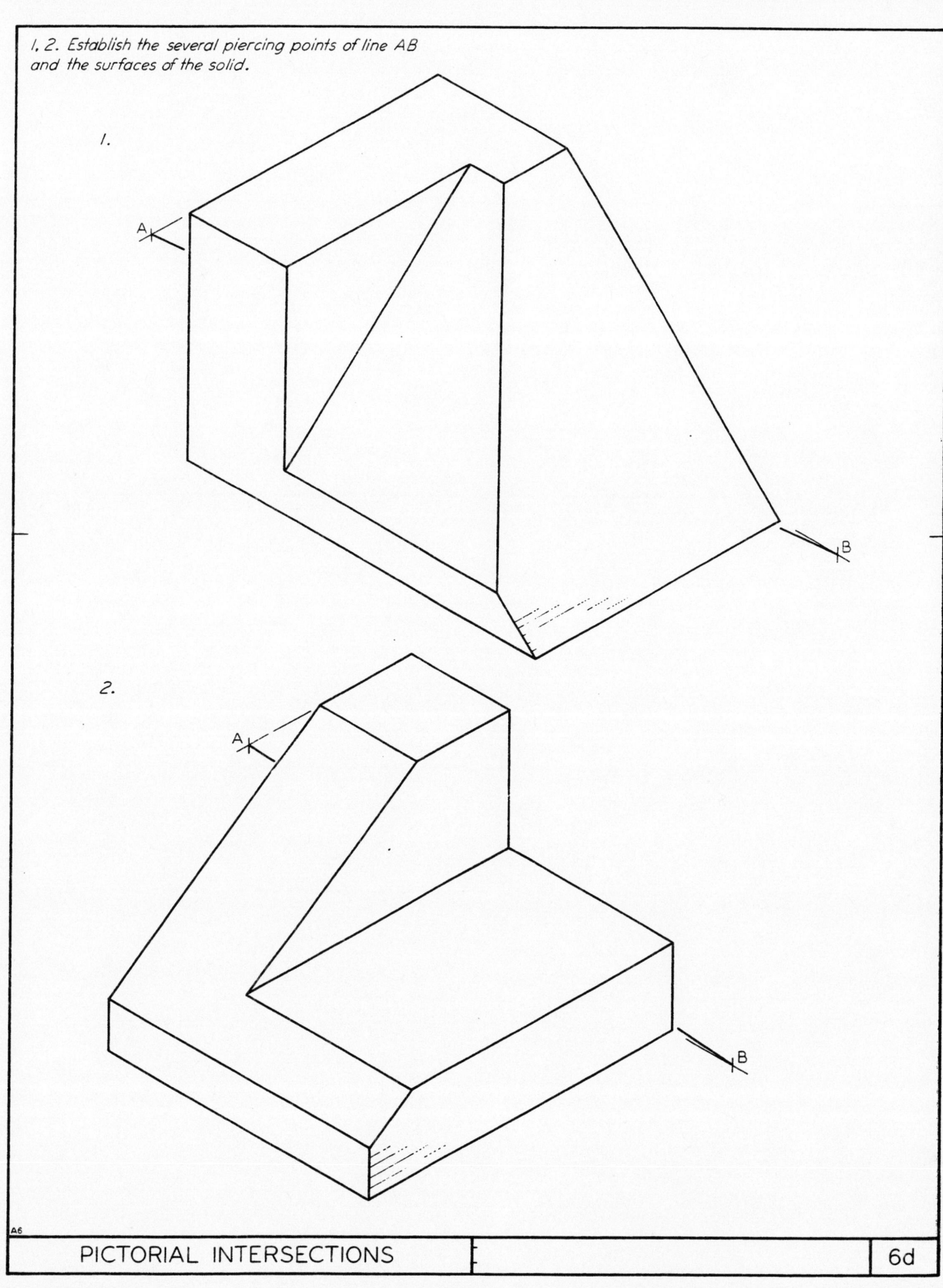

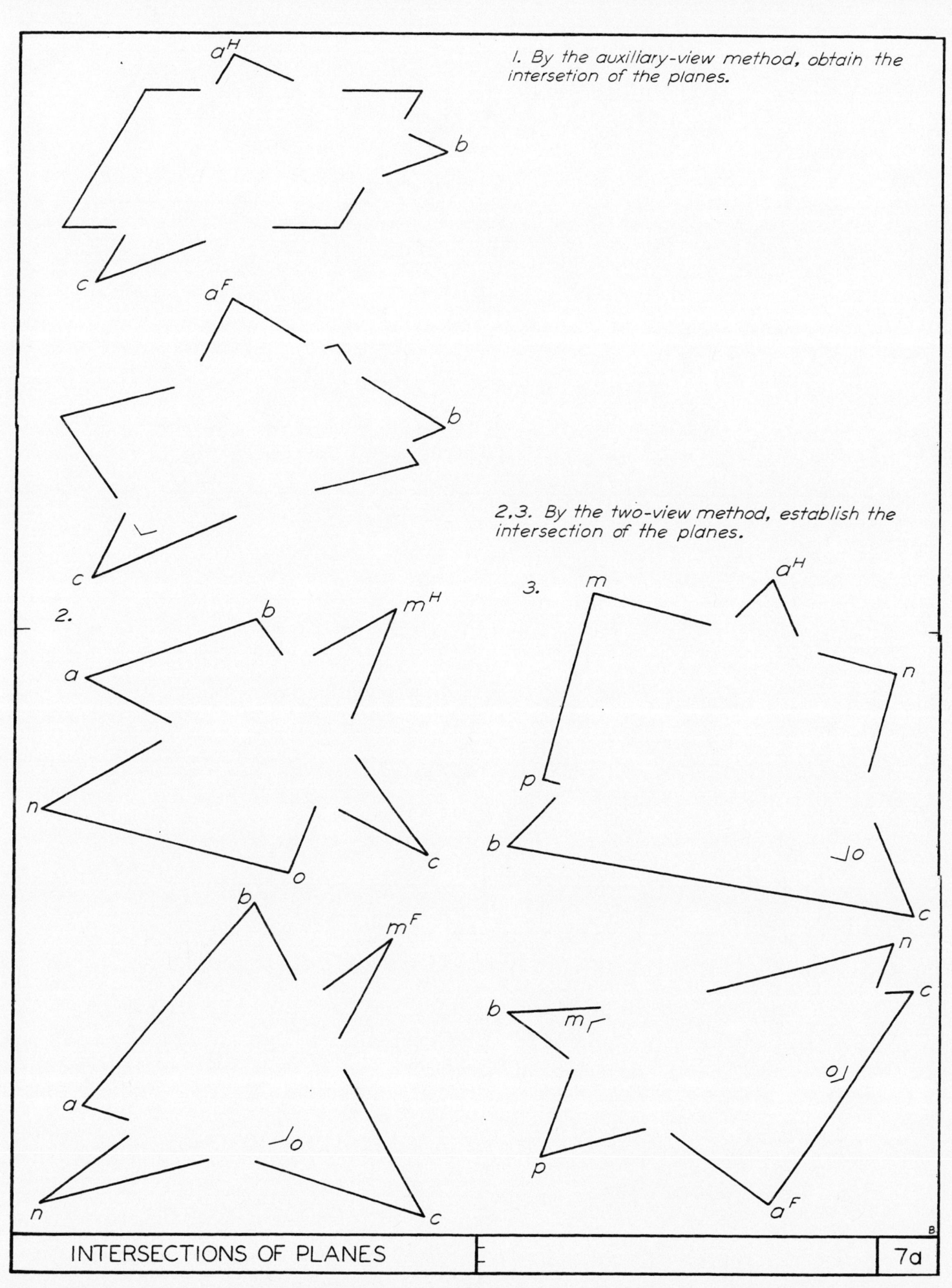

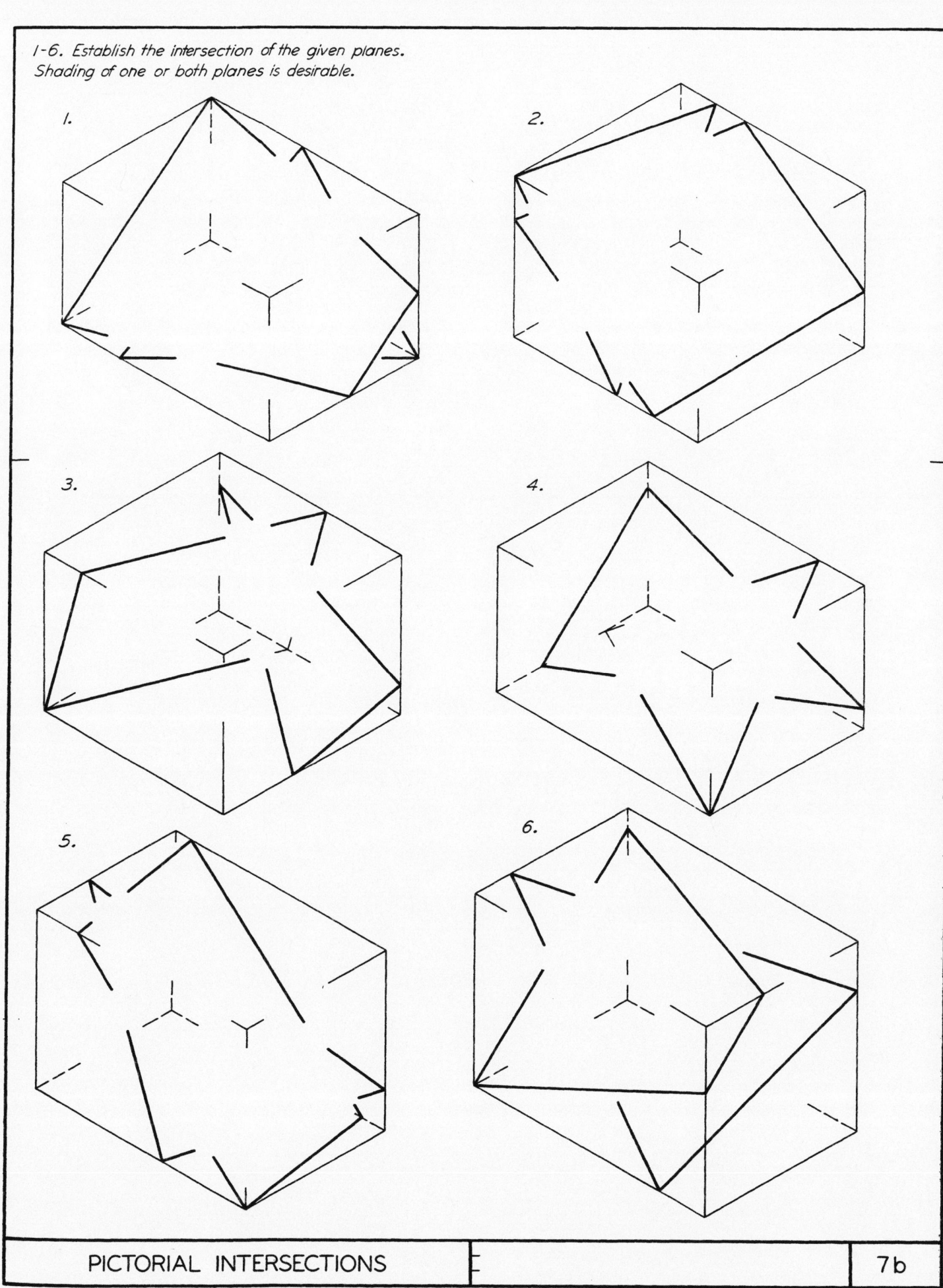

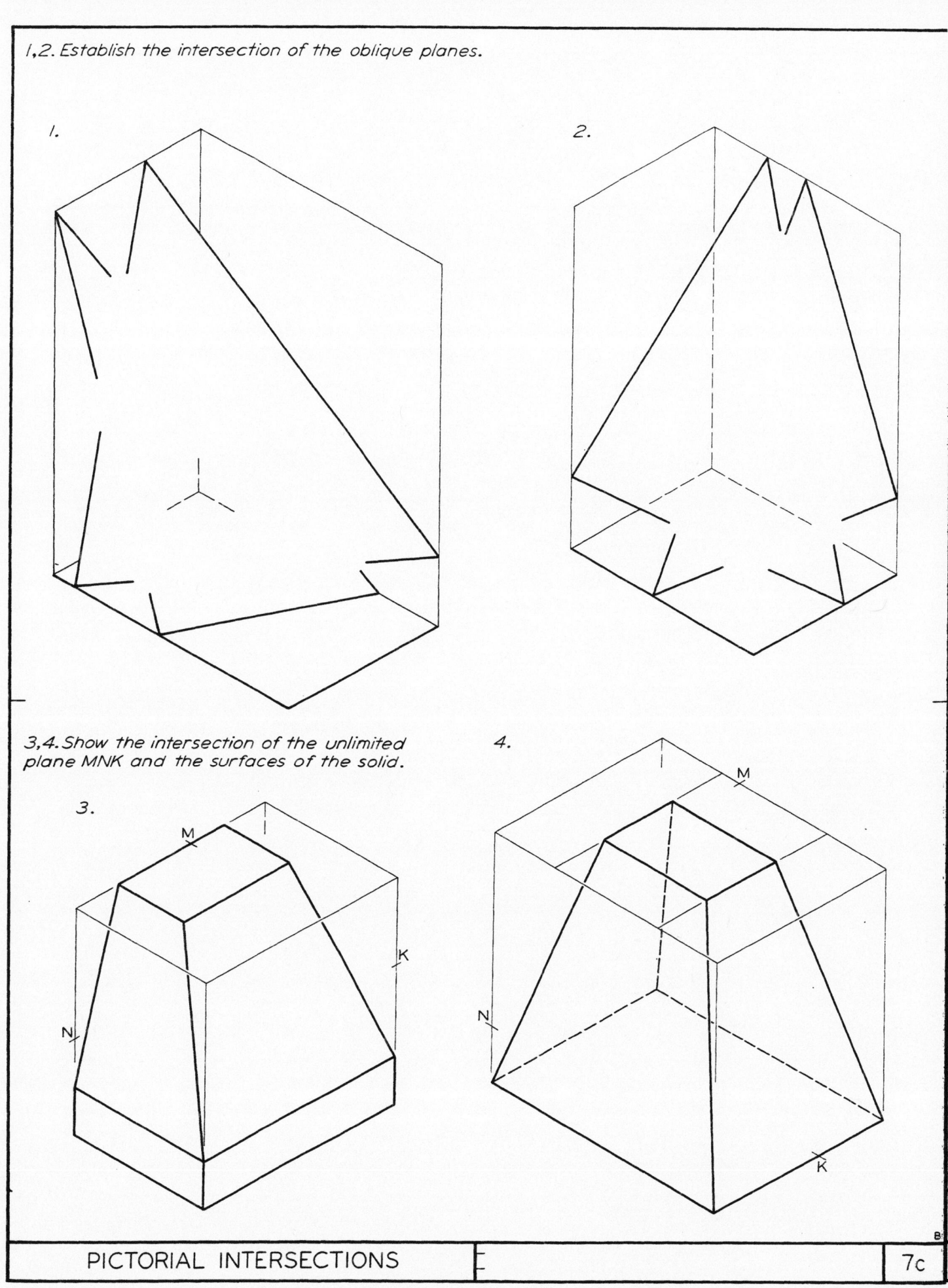

1. Determine the dihedral angles formed by the following planes of the tetrahedron:

VAB & ABC = _____
VAC & VAB = _____
VBC & VAC = _____

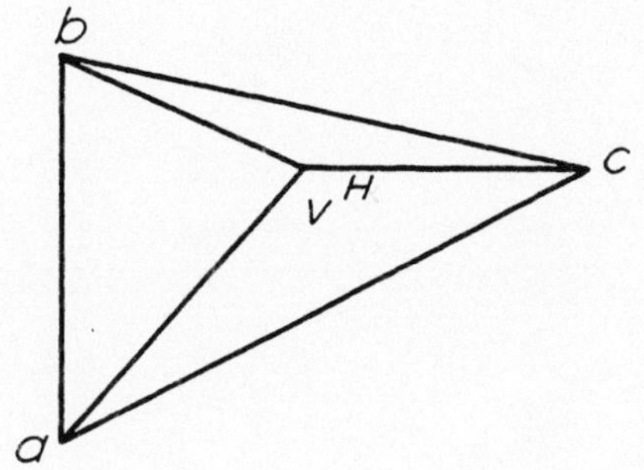

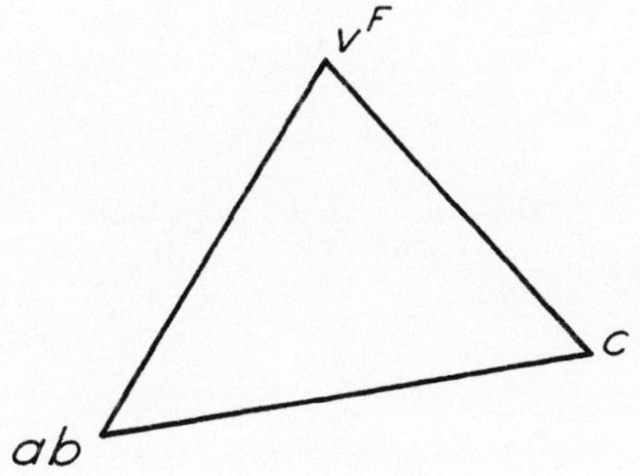

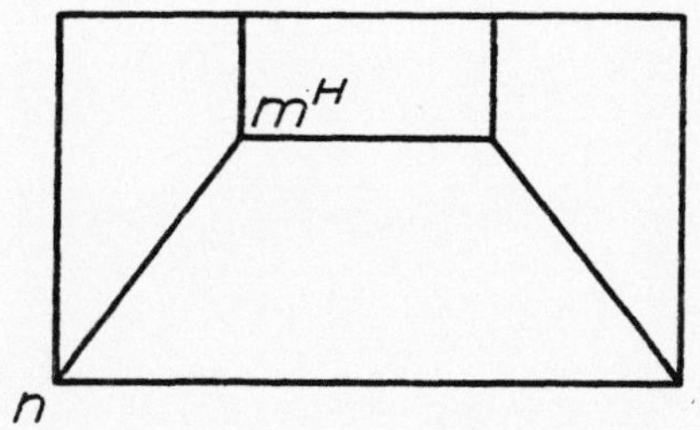

2. Obtain the angle for the bent-plate angle used along edge MN.

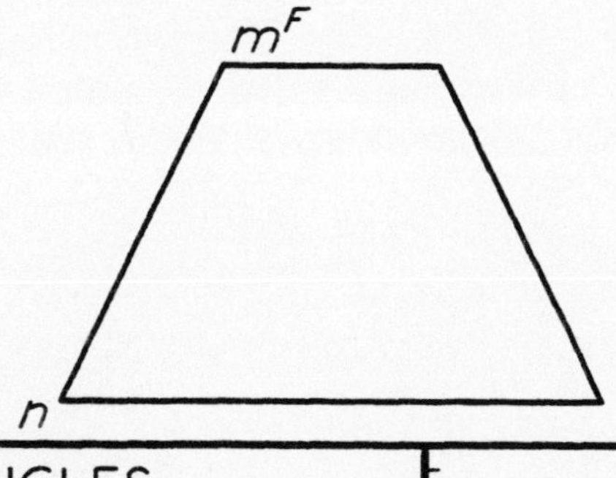

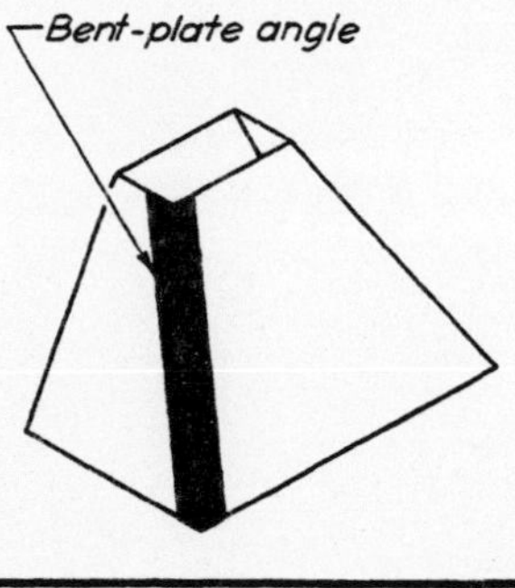

DIHEDRAL ANGLES 8a

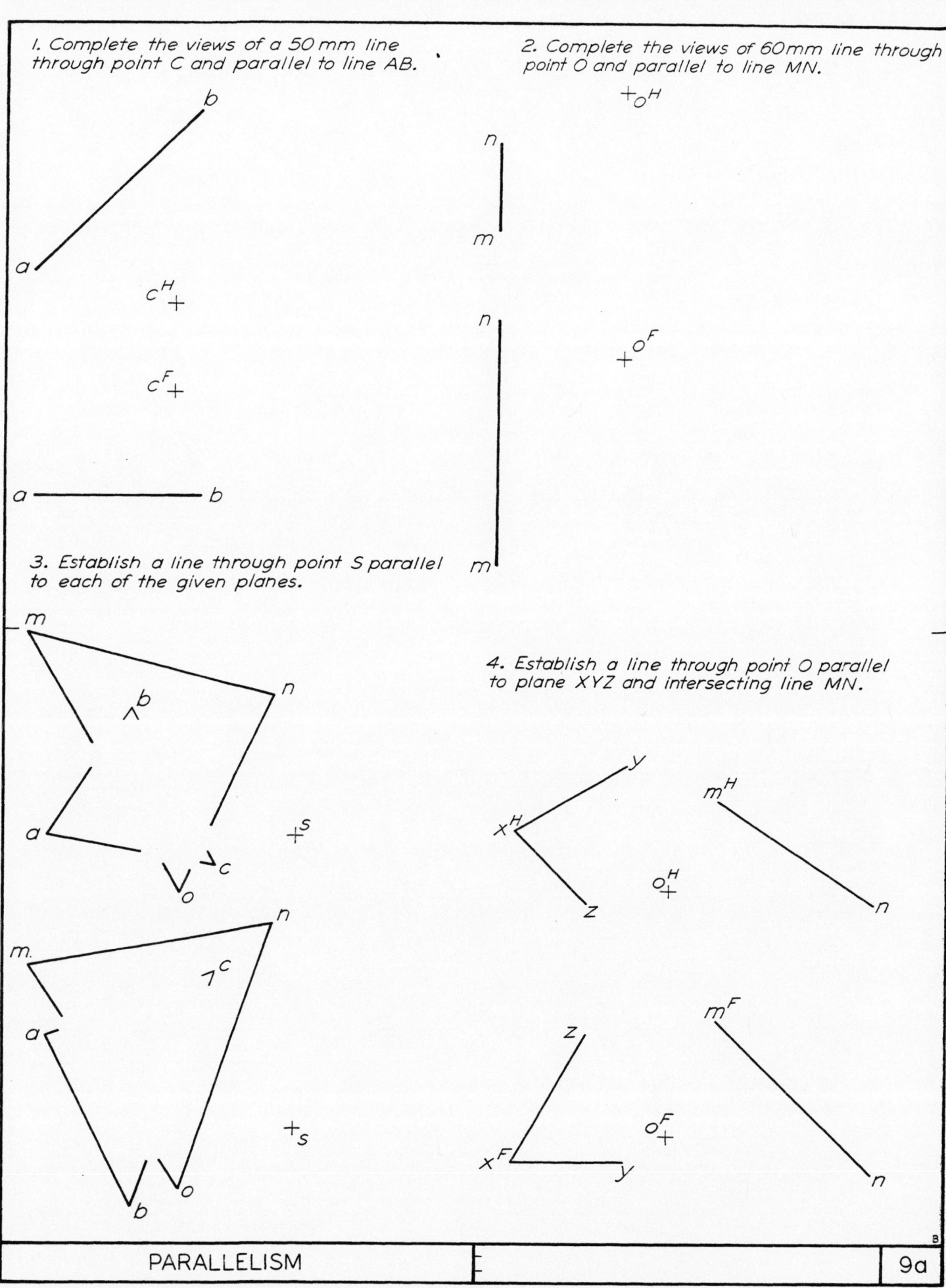

1. Specify which of the lines CE, OK, and MN are parallel to line AB.

Check (✓)	YES	NO
CE -		
MN -		
OK -		

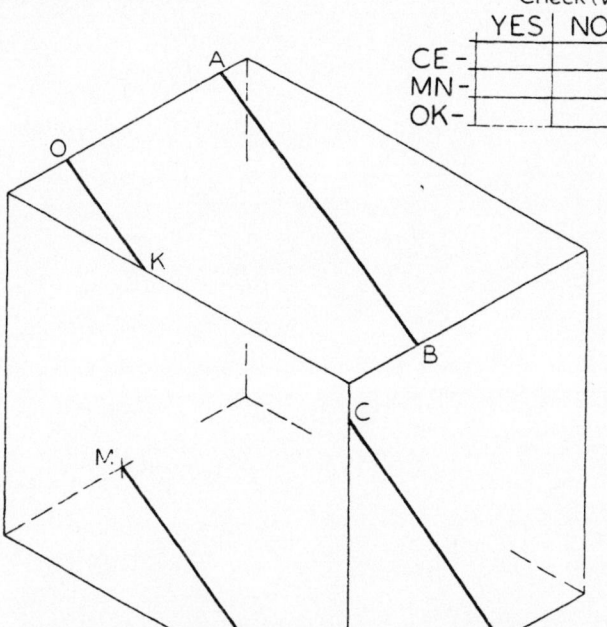

2. Establish a line on the surface 1, 2, 3, 4 through point K and parallel to plane MNO.

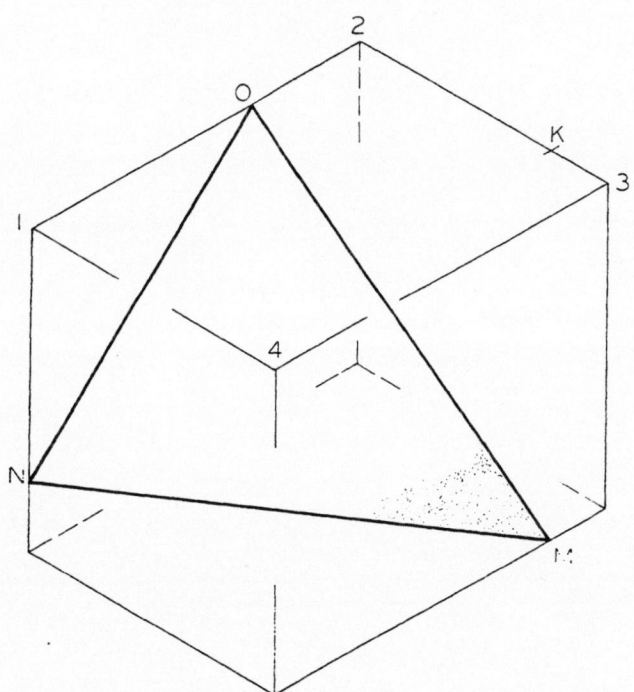

3. Establish a plane containing line MN and parallel to line CE.

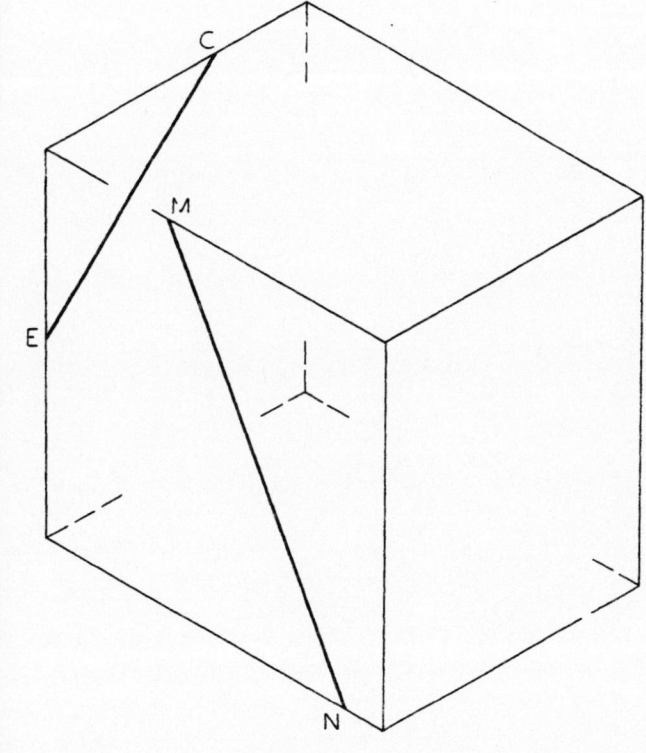

4. Locate a line through point O that intersects line MN and is parallel to plane ABC.

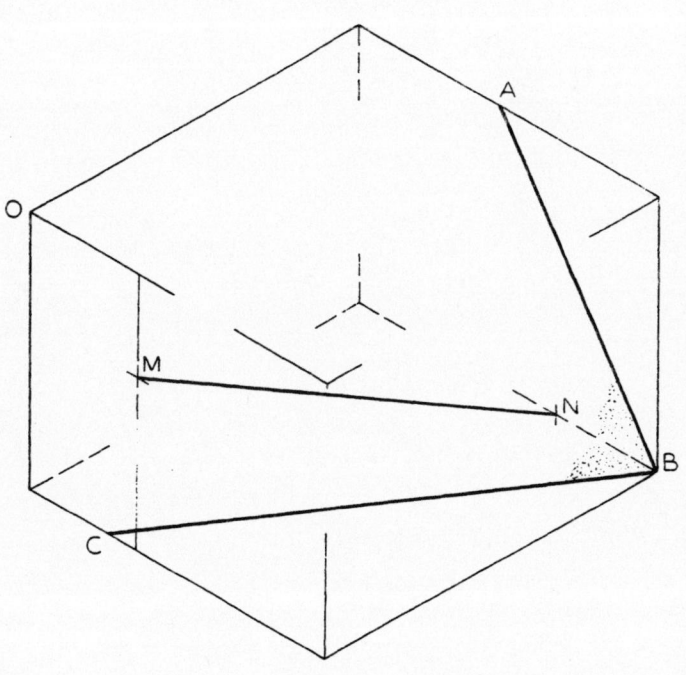

PICTORIAL PARALLELISM

1. Find the true length and the front and top views of the altitude of the pyramid with vertex at point V.

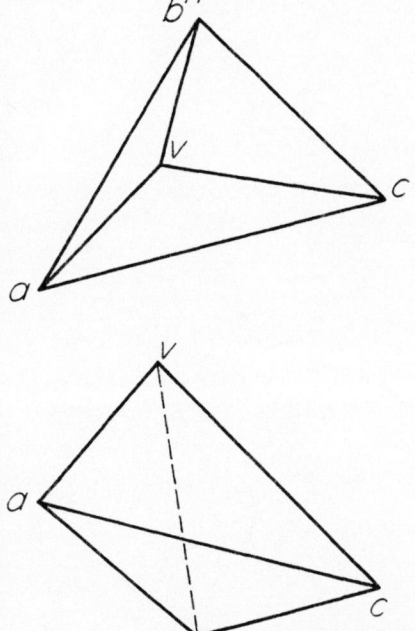

2. Obtain the orthographic projection of line MN on the given plane.

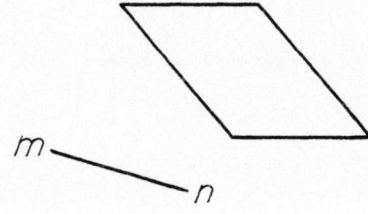

3. Locate the line from point O that is perpendicular to and intersects the given plane.

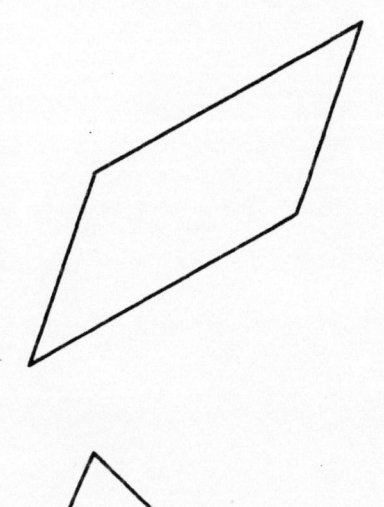

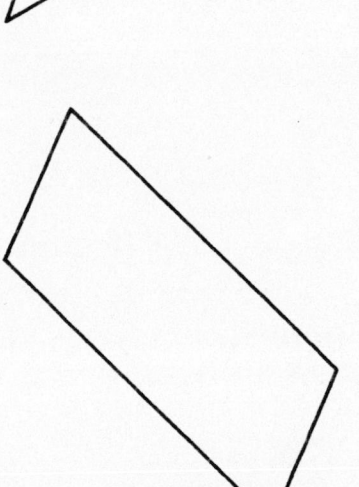

4. Establish the line from point C that is perpendicular to and intersects line MN.

PERPENDICULARITY 10a

1. Locate on line AB the center point for a circle that has line JK as a chord.

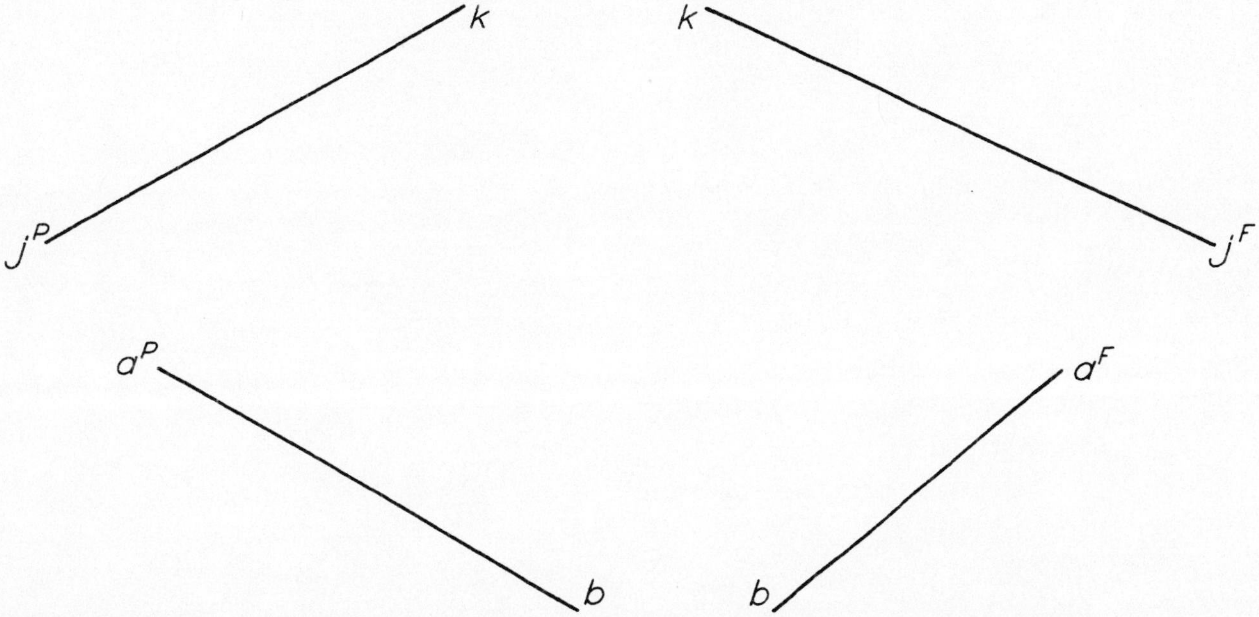

2. Establish the reflected light ray from the mirror surface ABCE for the light ray MN. Use only the given views.

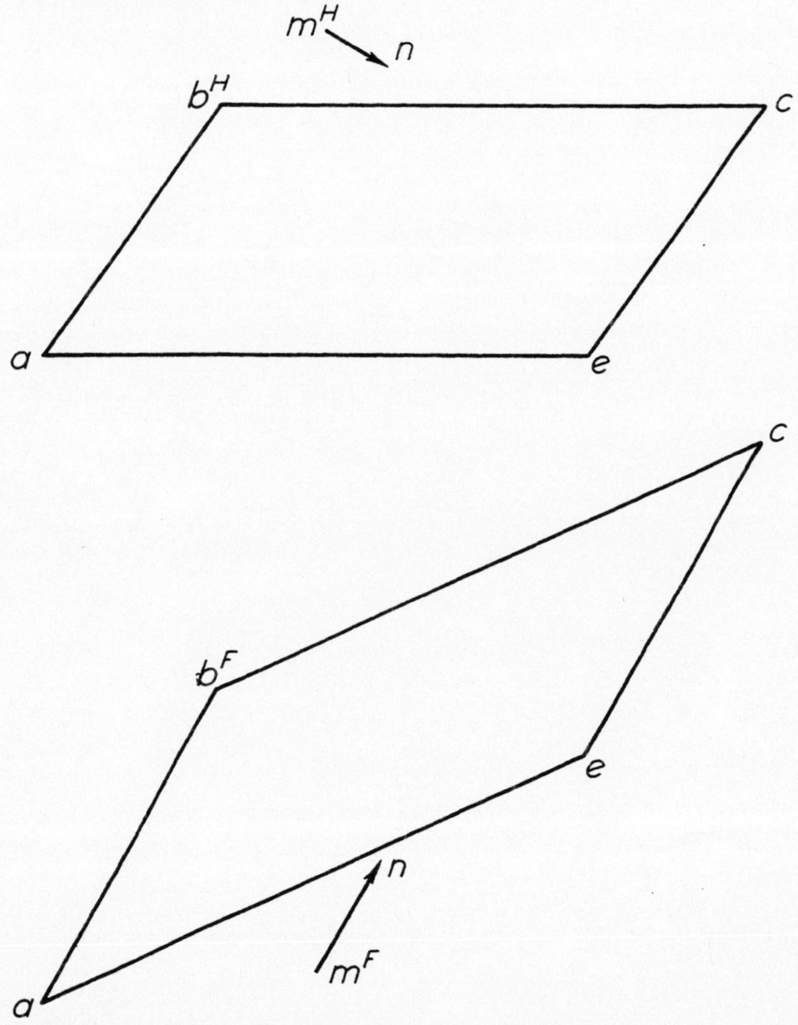

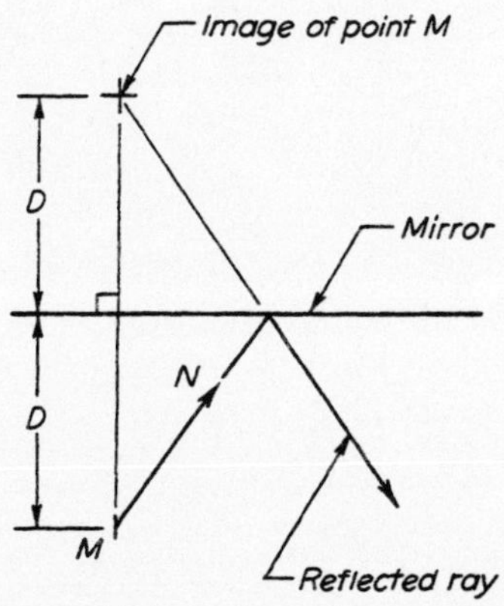

PERPENDICULARITY

10b

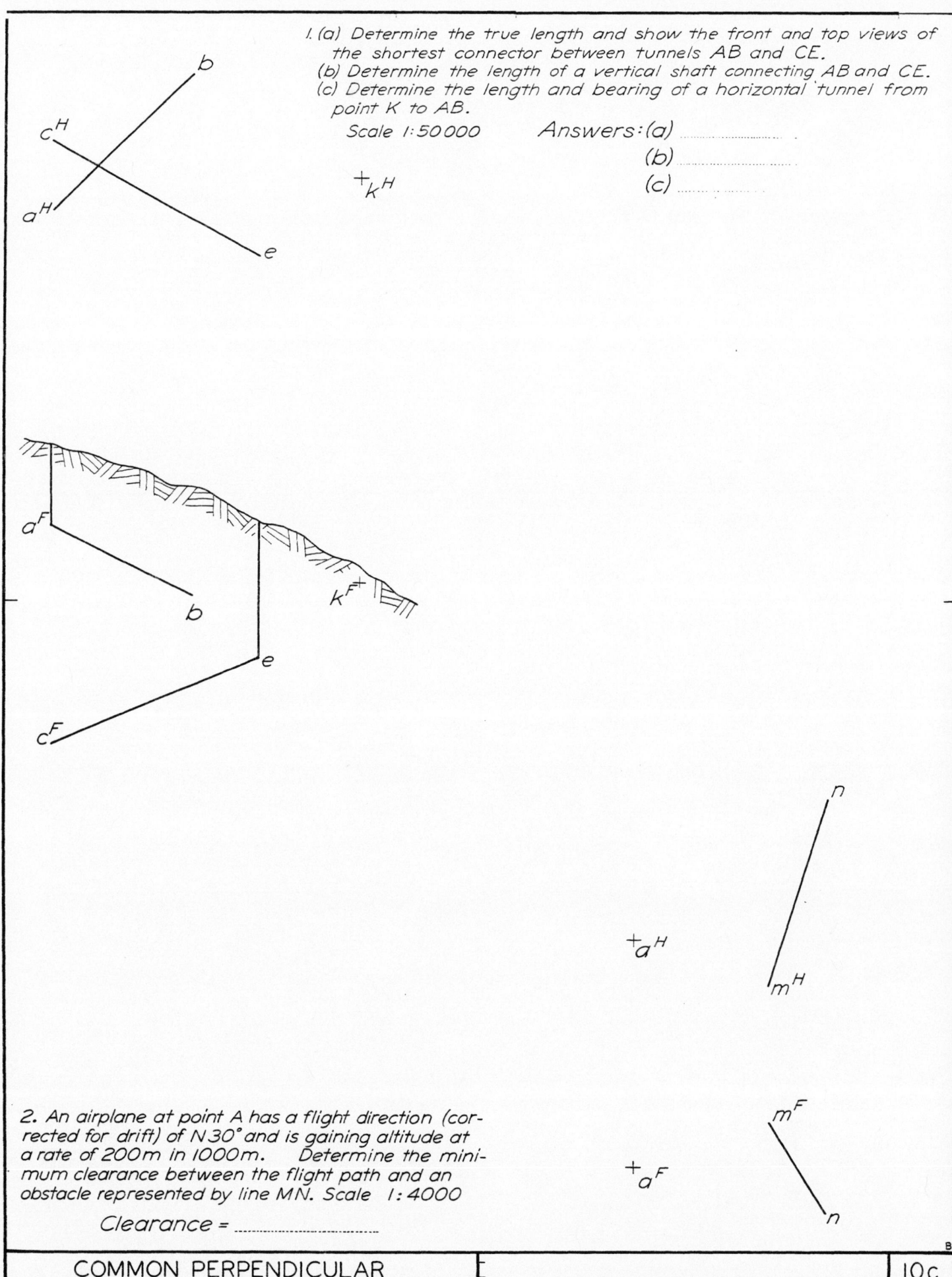

1. a. Obtain the angles that guy wire VA and mast VO make with roof plane MNOS.
   b. Obtain the angle that guy wire VB makes with roof plane MSKE.

   ANGLES -
   a. VO & MNOS - ....
      VA & MNOS - ....
   b. VB & MSK - ....

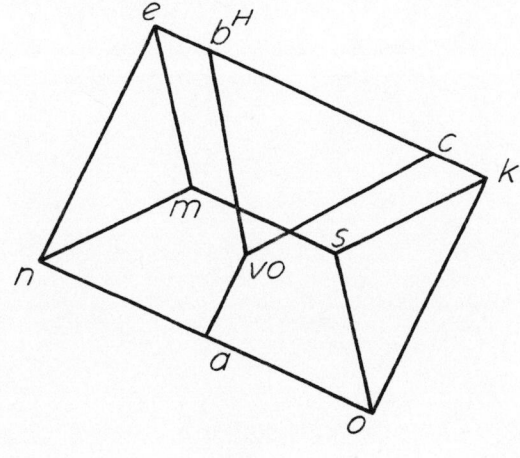

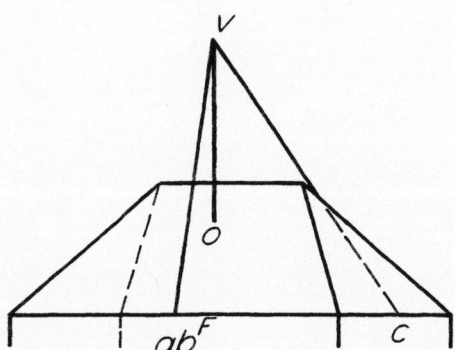

2. Obtain the angle of incidence of light ray MN and the mirror surface. Show the reflected ray in all views.

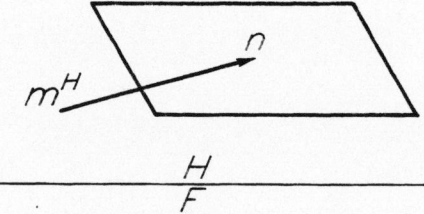

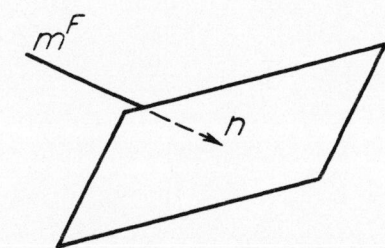

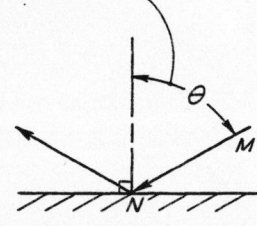

ANGLE - ....

ANGLE BETWEEN LINE & PLANE

1. Determine the strike and dip of the plane surface.
Strike = ...
Dip = ...

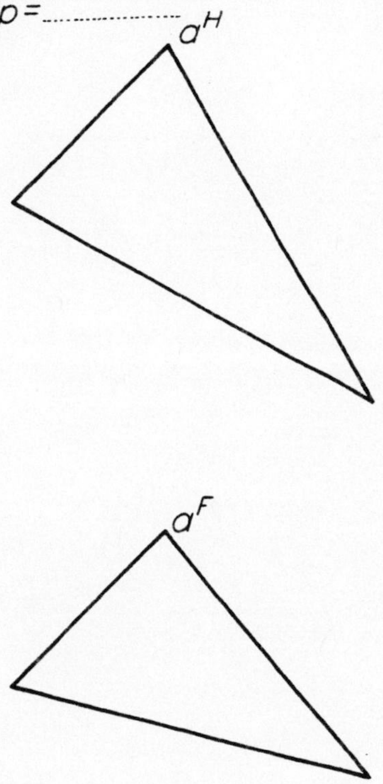

2. Establish the strike and dip of plane MON.
Strike = ...
Dip = ...

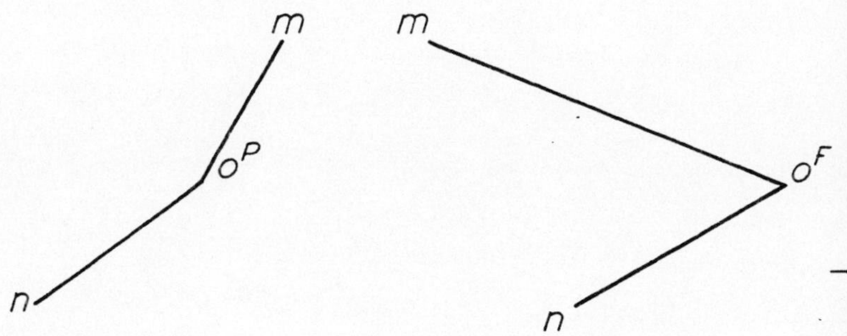

3. Points A, B, and C are on the upper bedding plane of a vein. Point E is on the parallel lower bedding plane. Find the strike, dip, and thickness of the vein.
Scale: 1/400 000.

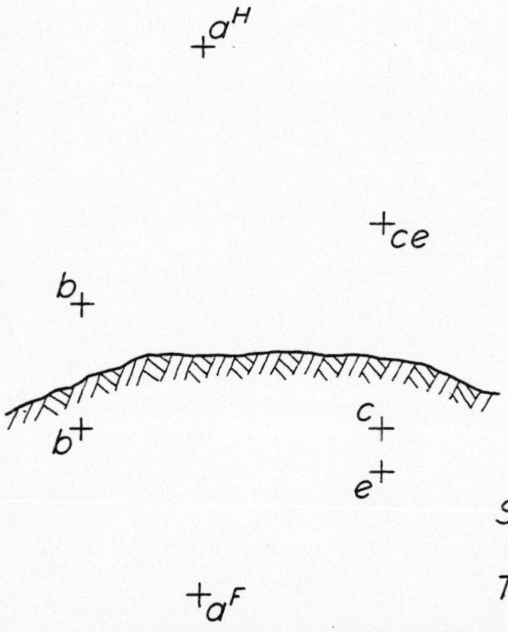

Strike = ...
Dip = ...
Thick. = ...

4. Points M and N lie in a plane that has a strike of N65° and a dip of 40°SE. Locate the top view of point N.

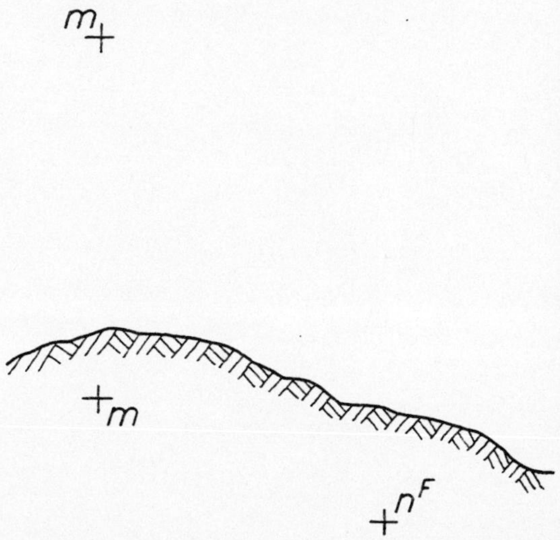

MINING AND GEOLOGY

12a

1. Points A, B, and C are on the upper outcrop line of an ore-vein and point M is on the lower outcrop line. Obtain the strike, dip, and thickness of the vein.
2. Plot contour lines for the adjacent area represented by the grid survey. Scale: 1/1000.
3. Establish the outcrop lines in both areas.

STRIKE = ... DIP = ... THICKNESS = ...

CONTOUR MAP AND OUTCROP   12b

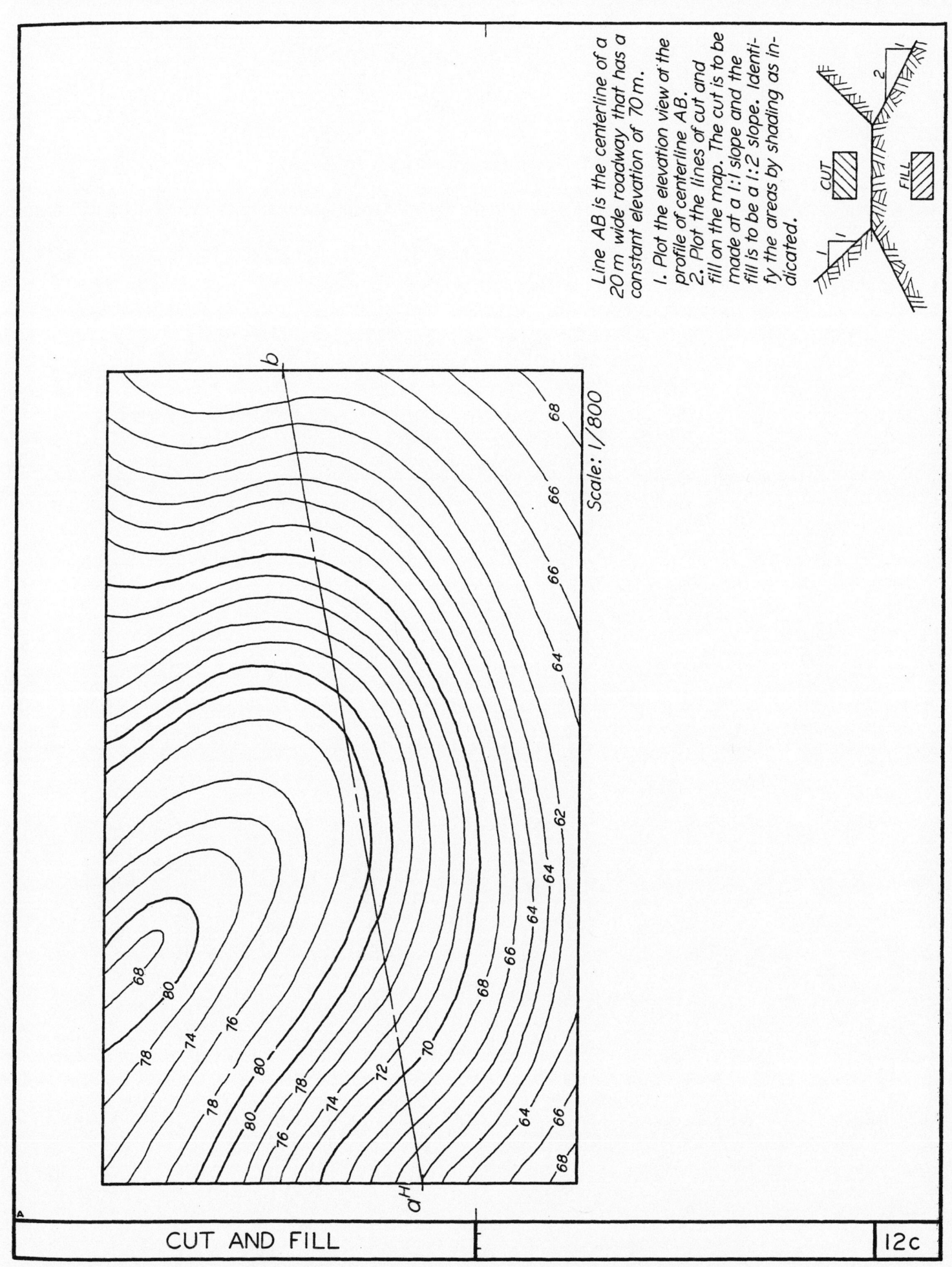

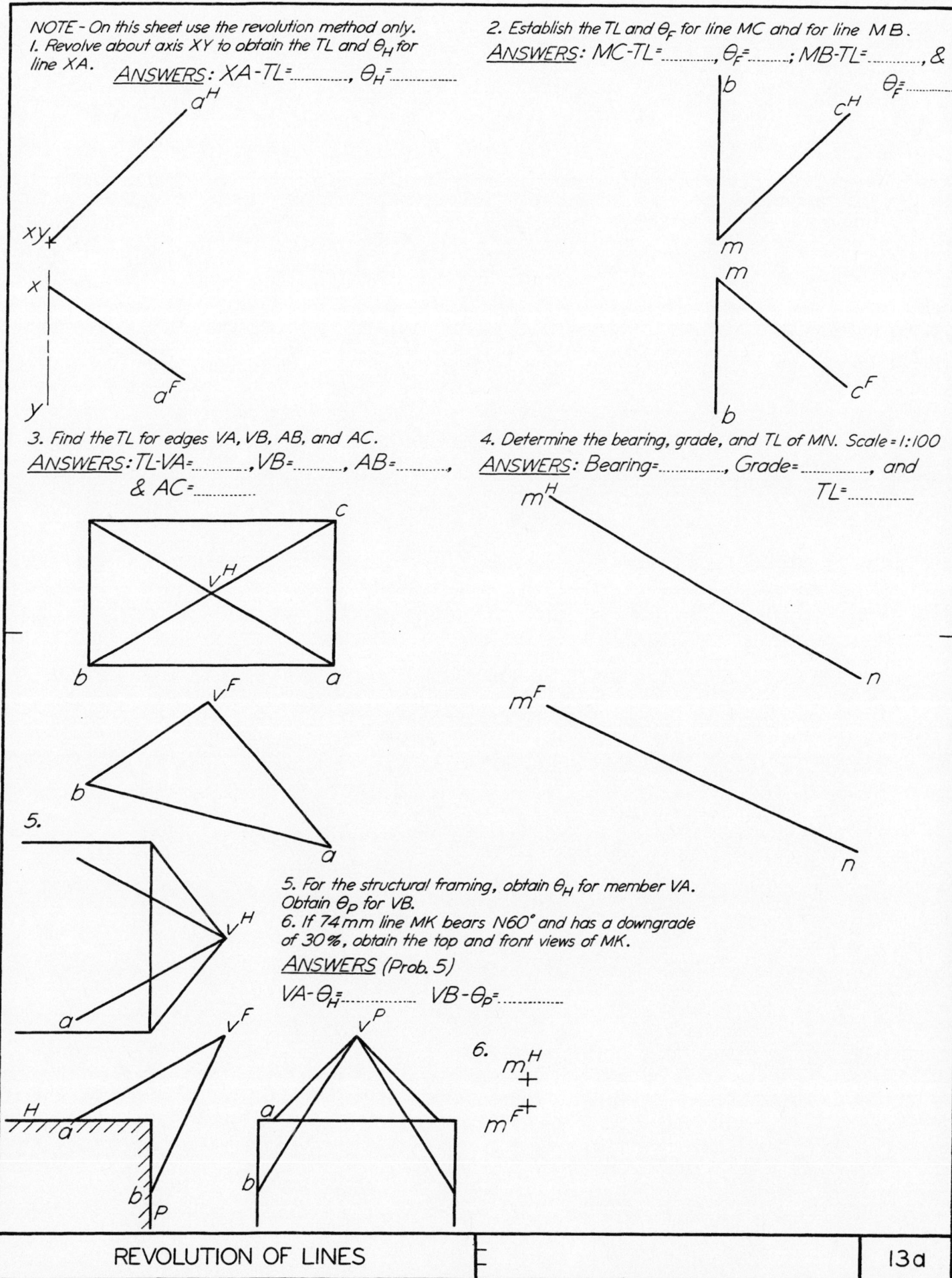

1. Determine the magnitude of the resultant of the three coplanar forces. Lay out the vectors in alphabetical order. Scale 1mm=20N

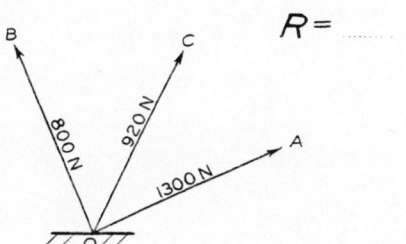

R=

2. Find the magnitude of the resultant of the coplanar forces acting on a 320 kg anchorage.
   Will the anchorage be lifted by the given forces?
   If the coefficient of friction is 0.5, will the anchorage have a tendency to slide?
   Scale 1mm=20 N

R=

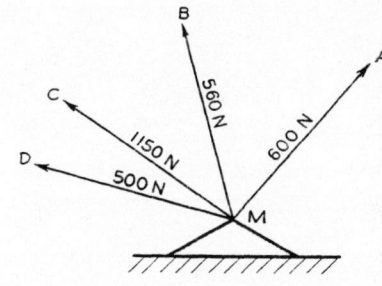

o_+^F

3. A ship at G is following a course of N-345° at 18 knots. A ship at K is following a course of N-60° at 21 knots. How close will the ships pass? How much time will elapse before this situation occurs?
   Vector scale 1cm=4 knots
   Distance scale 1/10 000

Distance=
   Time=

$+m^F$

k_+^H

$+g^H$

COPLANAR VECTORS

14a

1. Find the H and F projections and the true magnitude of the resultant of the three forces A, B, and C. Complete the parallelepiped using line weights similar to those in the pictorial.
Scale: 1mm / 10N.

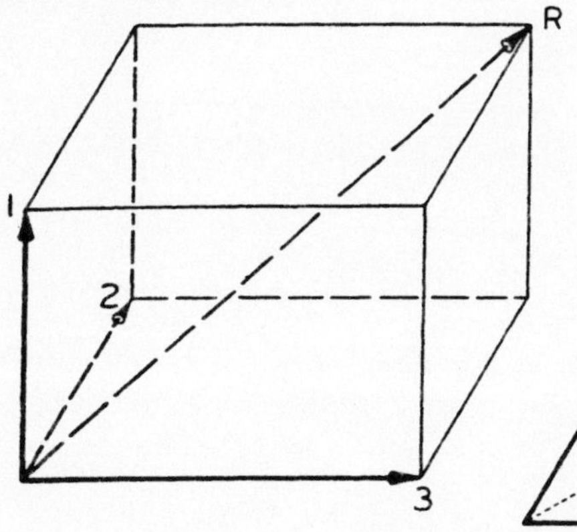

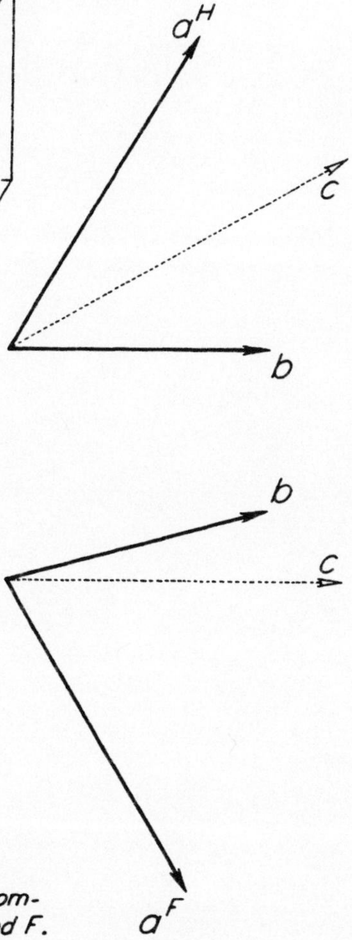

R = ...

2. Resolve the given vector R into components along the members D, E, and F.

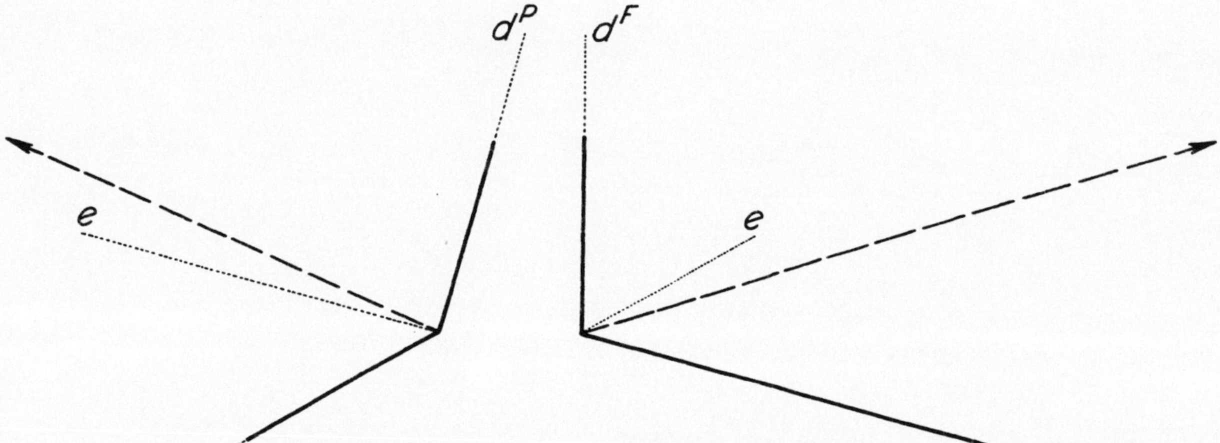

NON-COPLANAR VECTORS

14b

1. Find the magnitude and sense of the resultant (R) of the four vectors A, B, C, and D. Start the Space Diagram at point X.   Scale: 1mm = 10 N

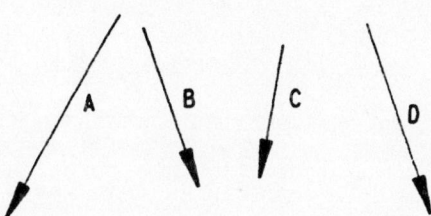

+ X

2. Find the Reactions R_L and R_R at the Beam loaded as shown.   Space Scale: 1:100. Vector Scale: 1m = 20 N.

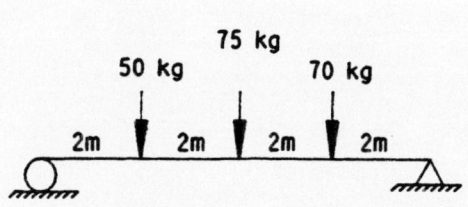

+ X

O +

Non Concurrent Vectors | 14C_A

1. Find the Reactions R_L and R_R of the Truss. Find the loads in each member of the Truss. Locate the start of the Stress Diagram and Vector Diagram at X. Scales:

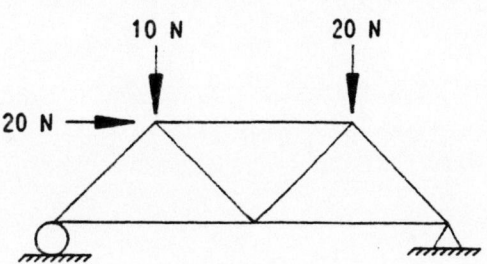

2.   Scale:

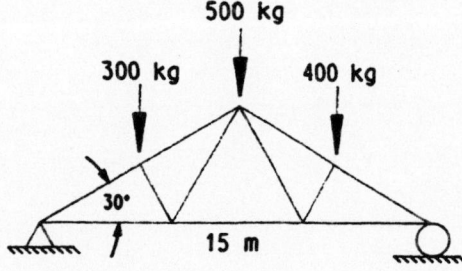

Truss Stresses

14D$_A$

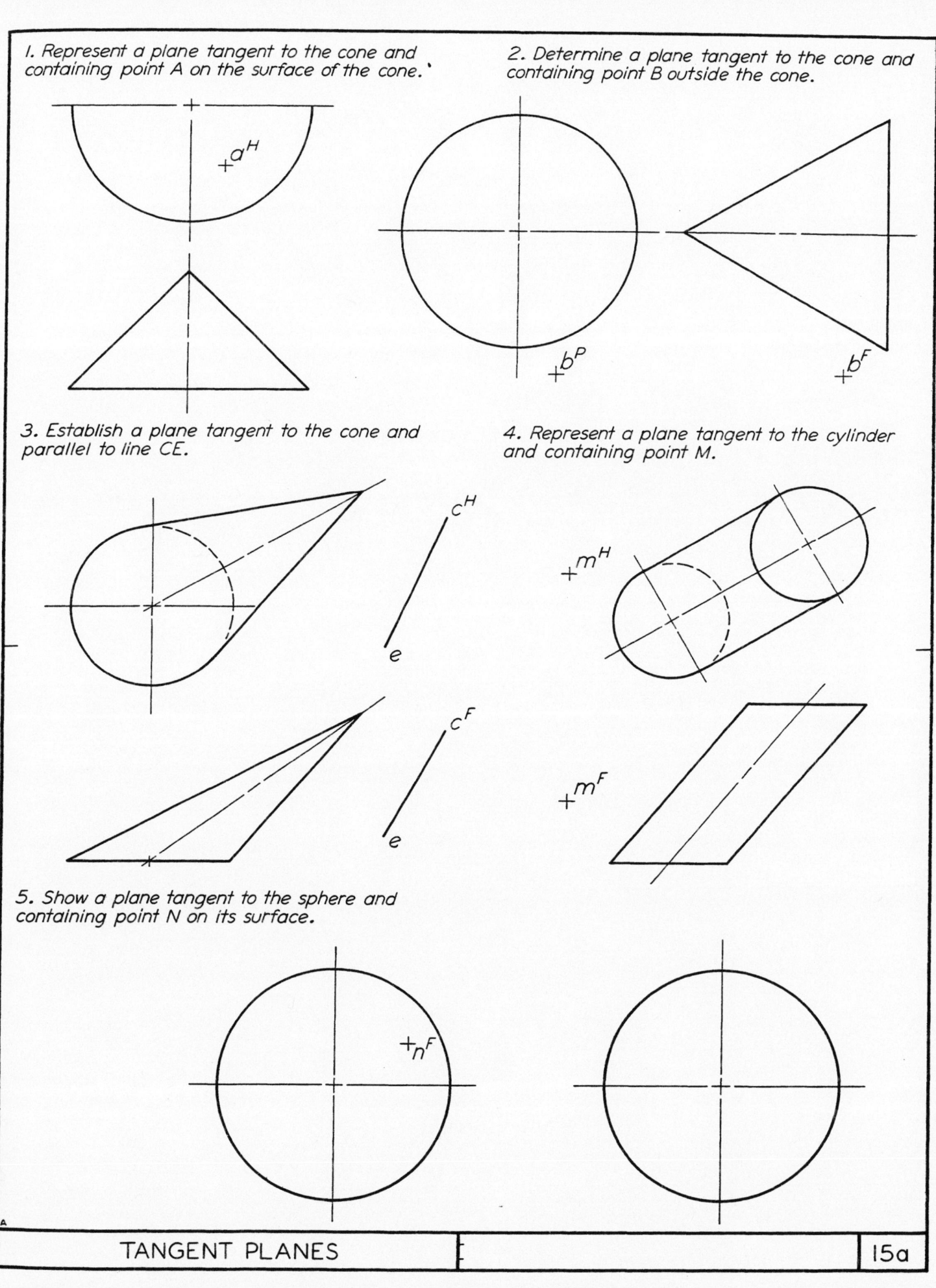

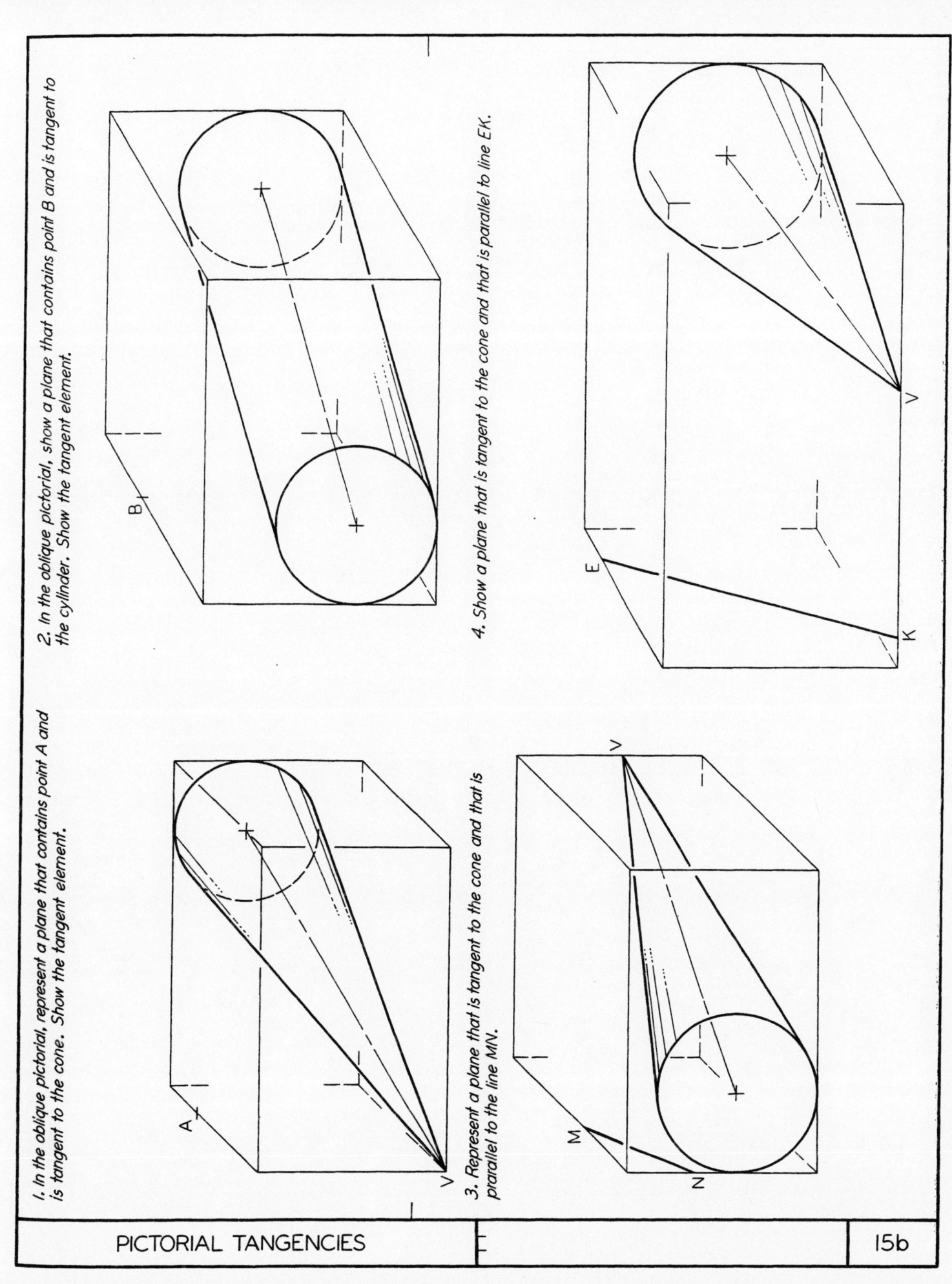

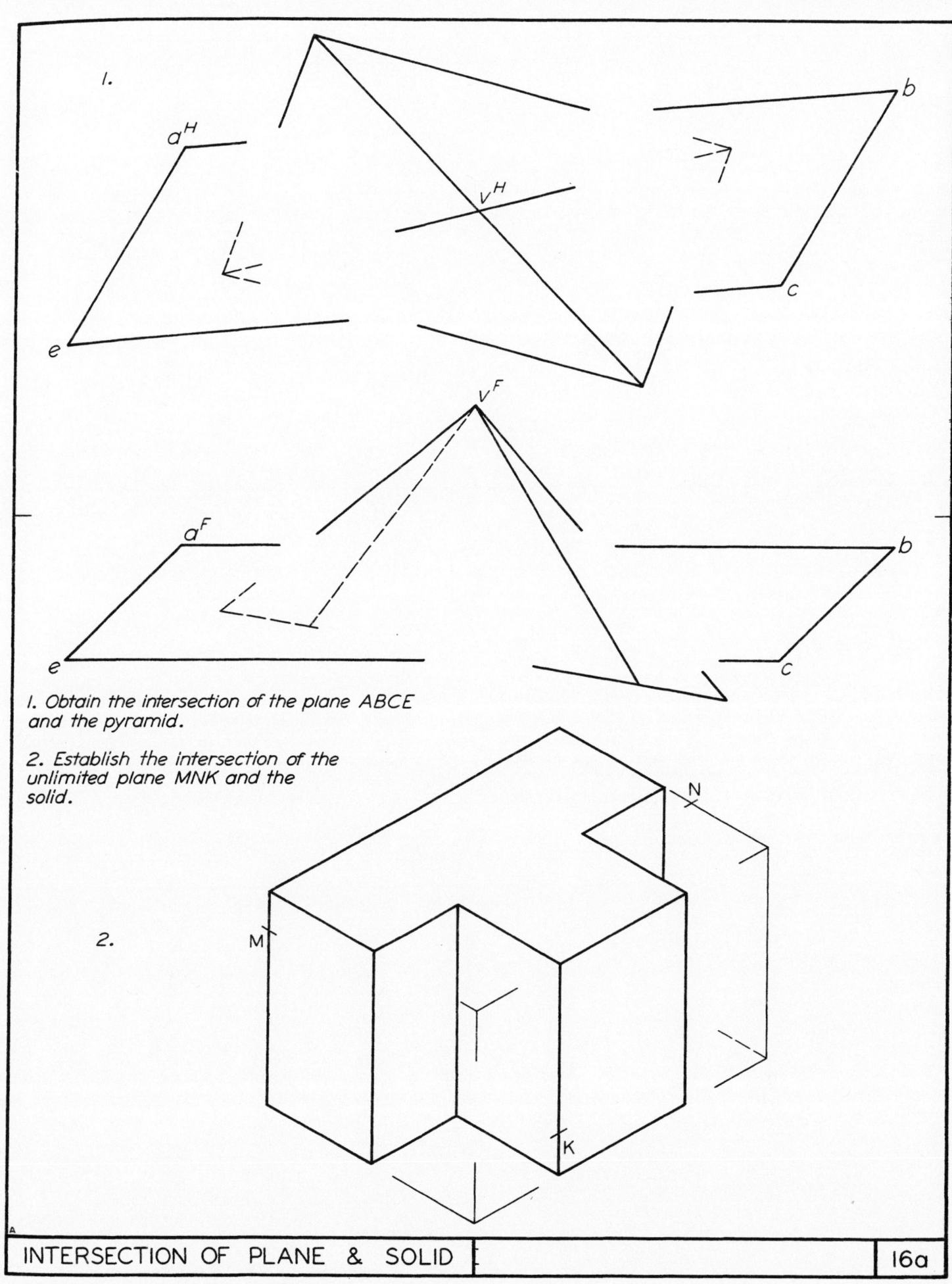

1. Obtain the intersection of the plane ABCE and the pyramid.

2. Establish the intersection of the unlimited plane MNK and the solid.

INTERSECTION OF PLANE & SOLID — 16a

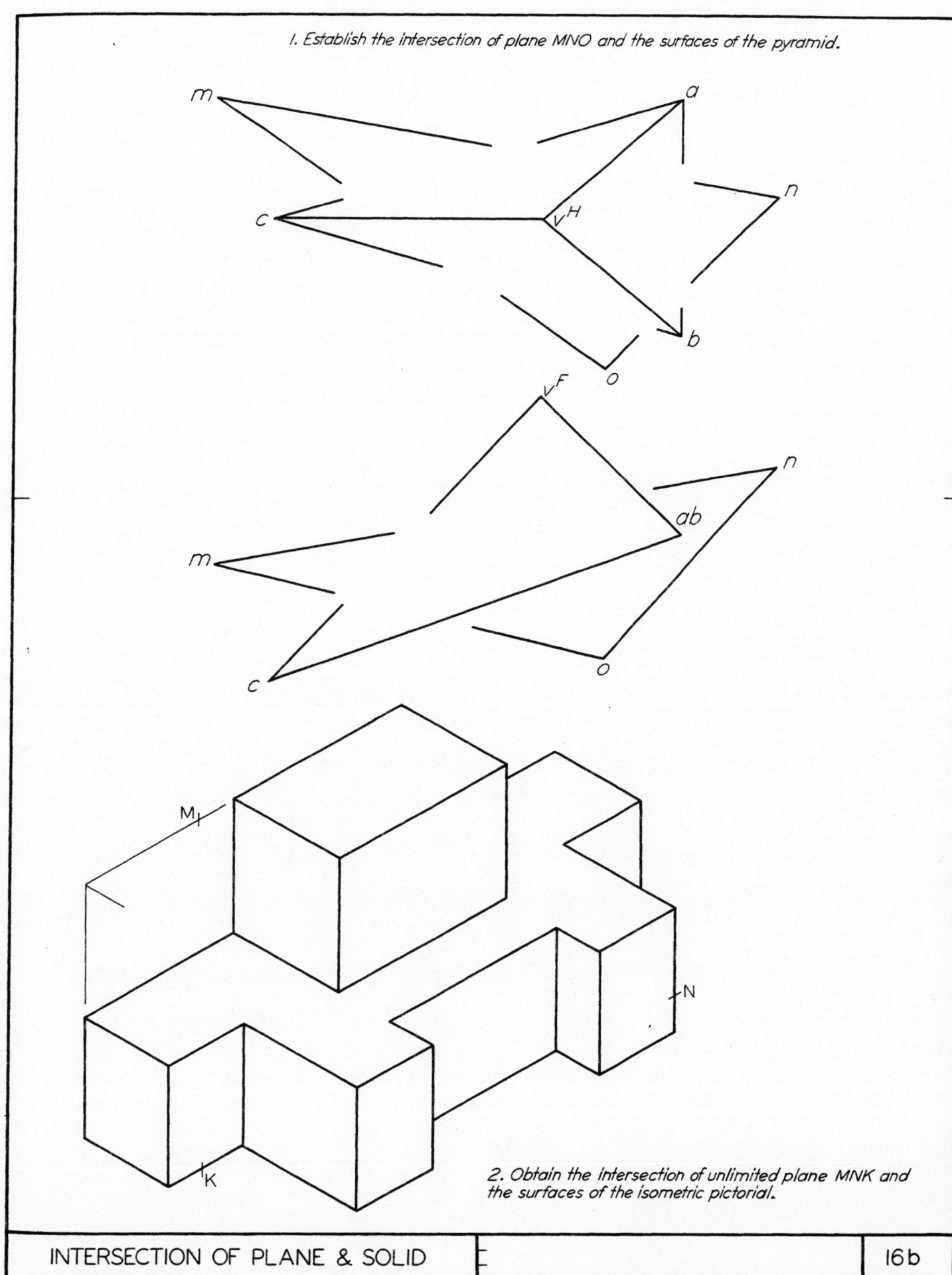

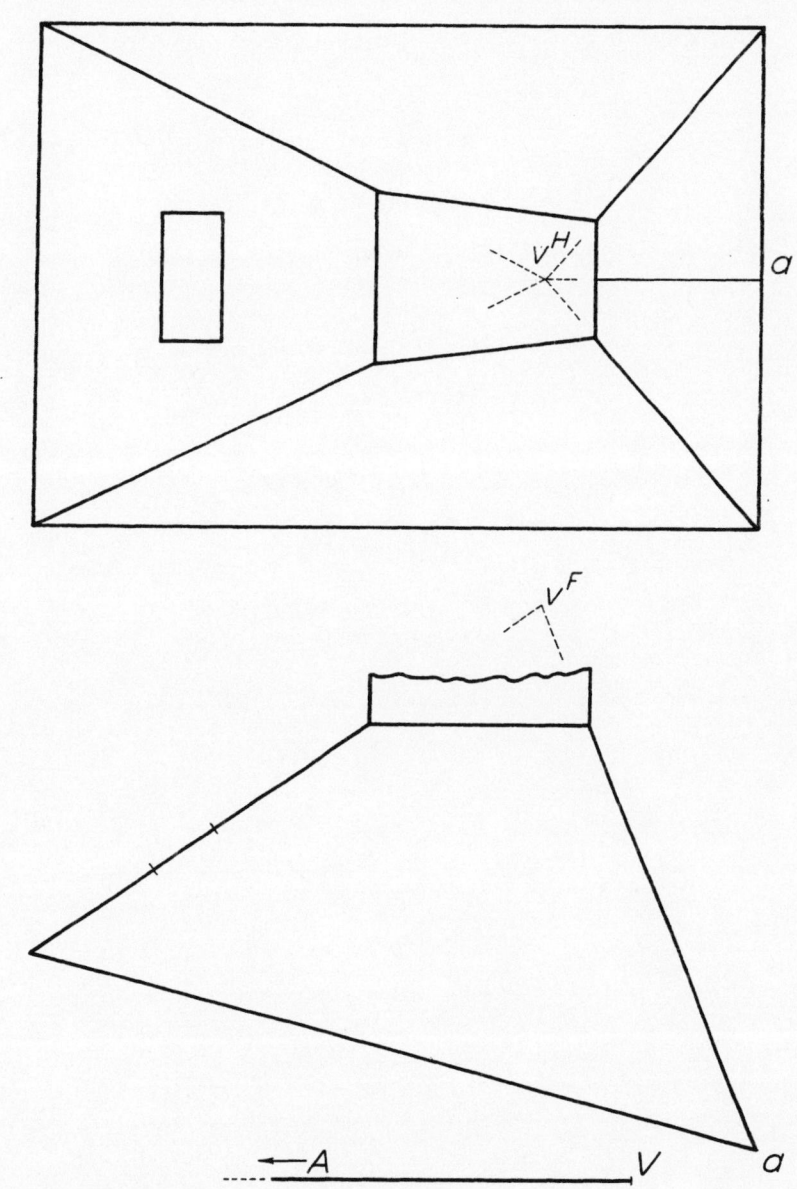

Provide a half development of the
pyramidal portion of the transition unit.
Start the layout along the convenient
dividing edge VA.

RADIAL LINE DEVELOPMENT   17a

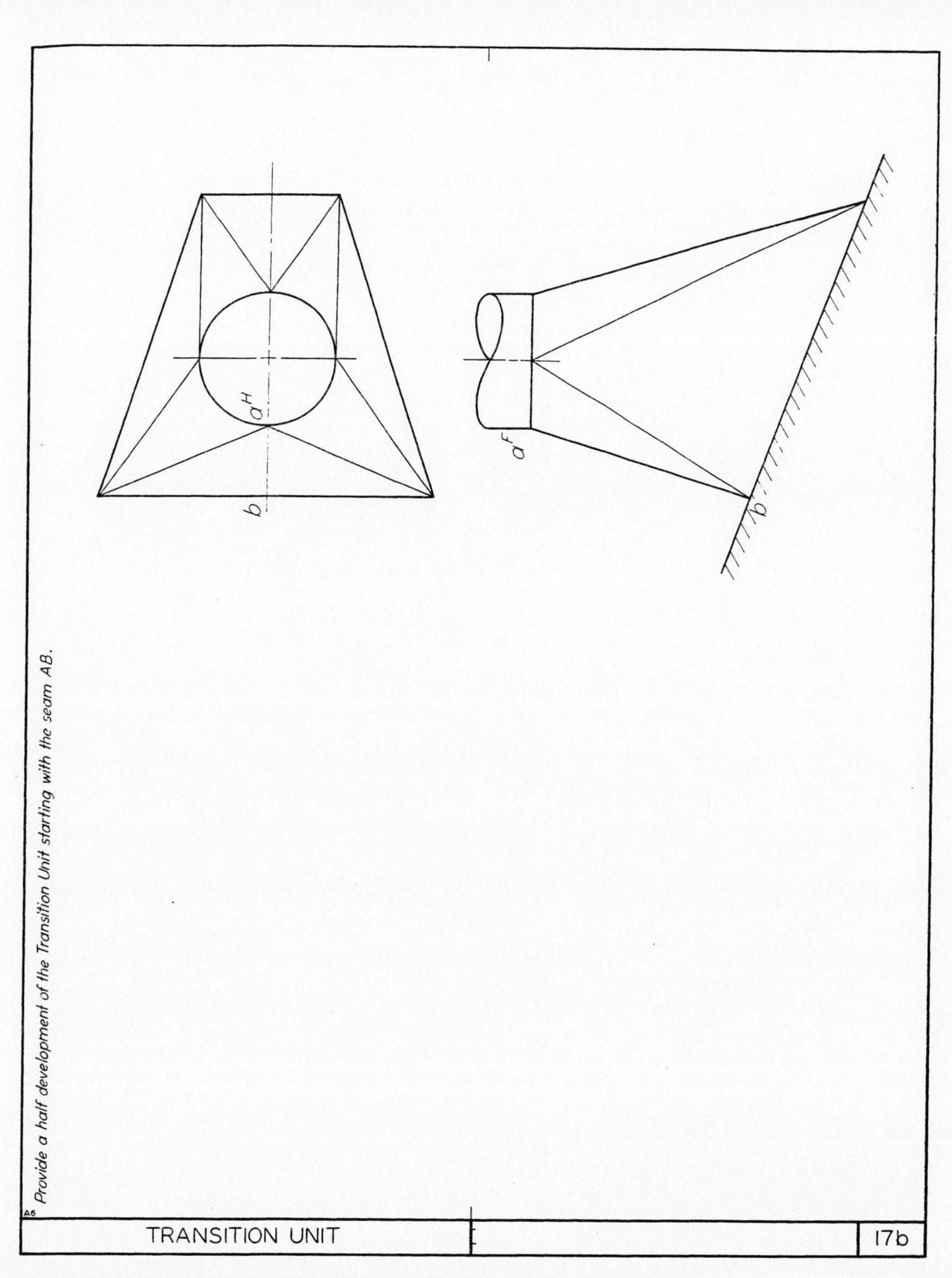

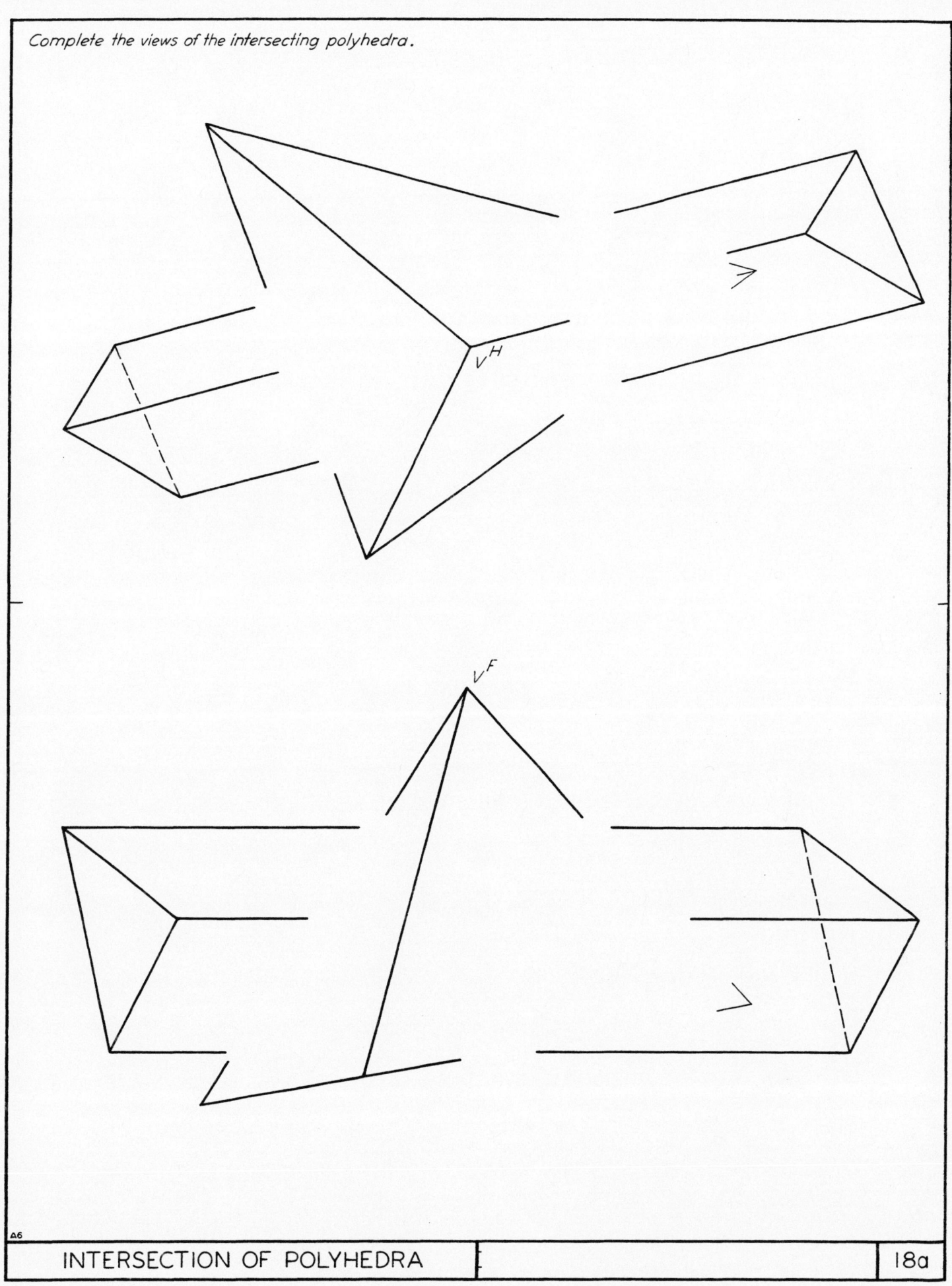

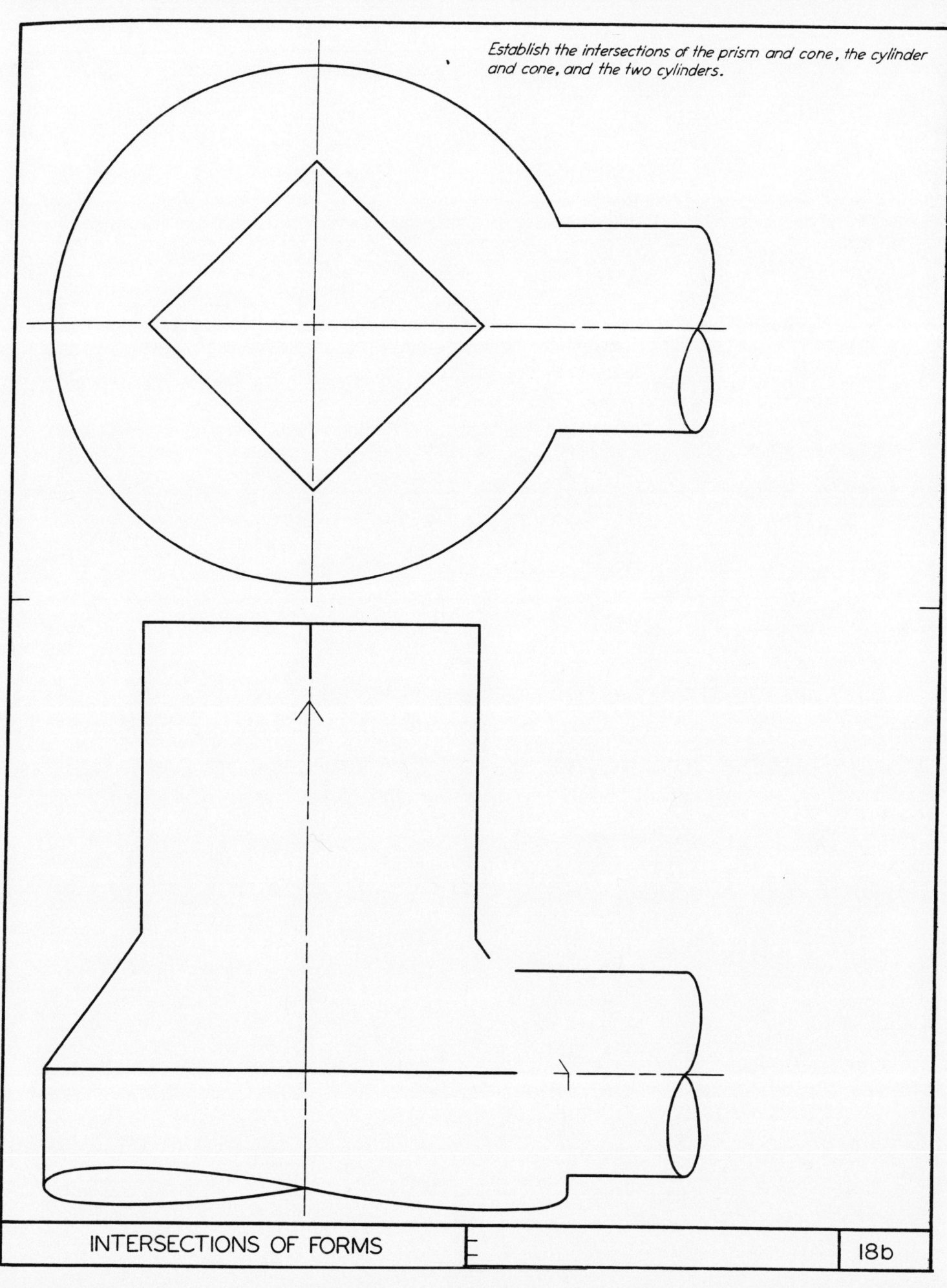

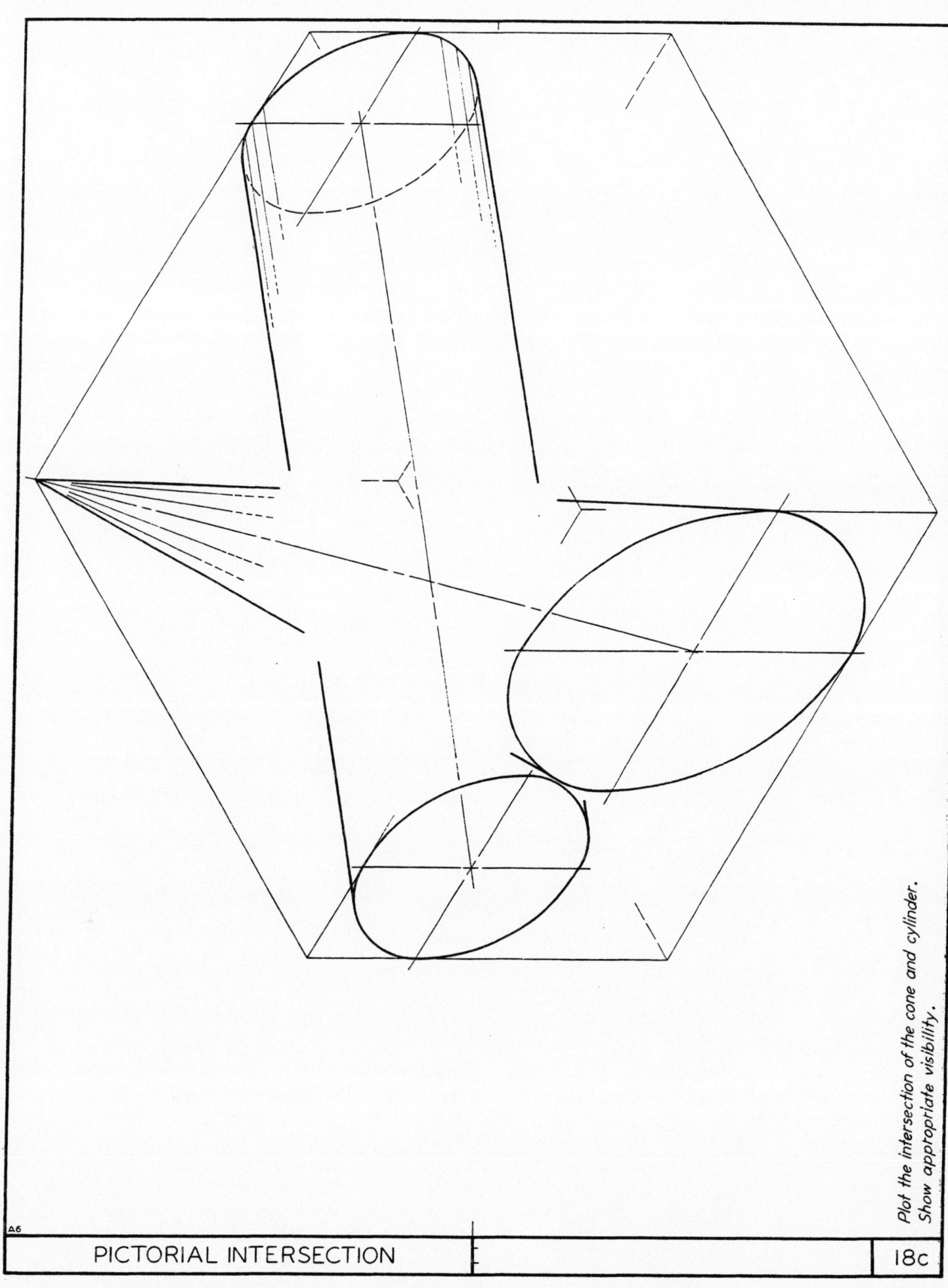

PICTORIAL INTERSECTION — 18c

Plot the intersection of the cone and cylinder. Show appropriate visibility.

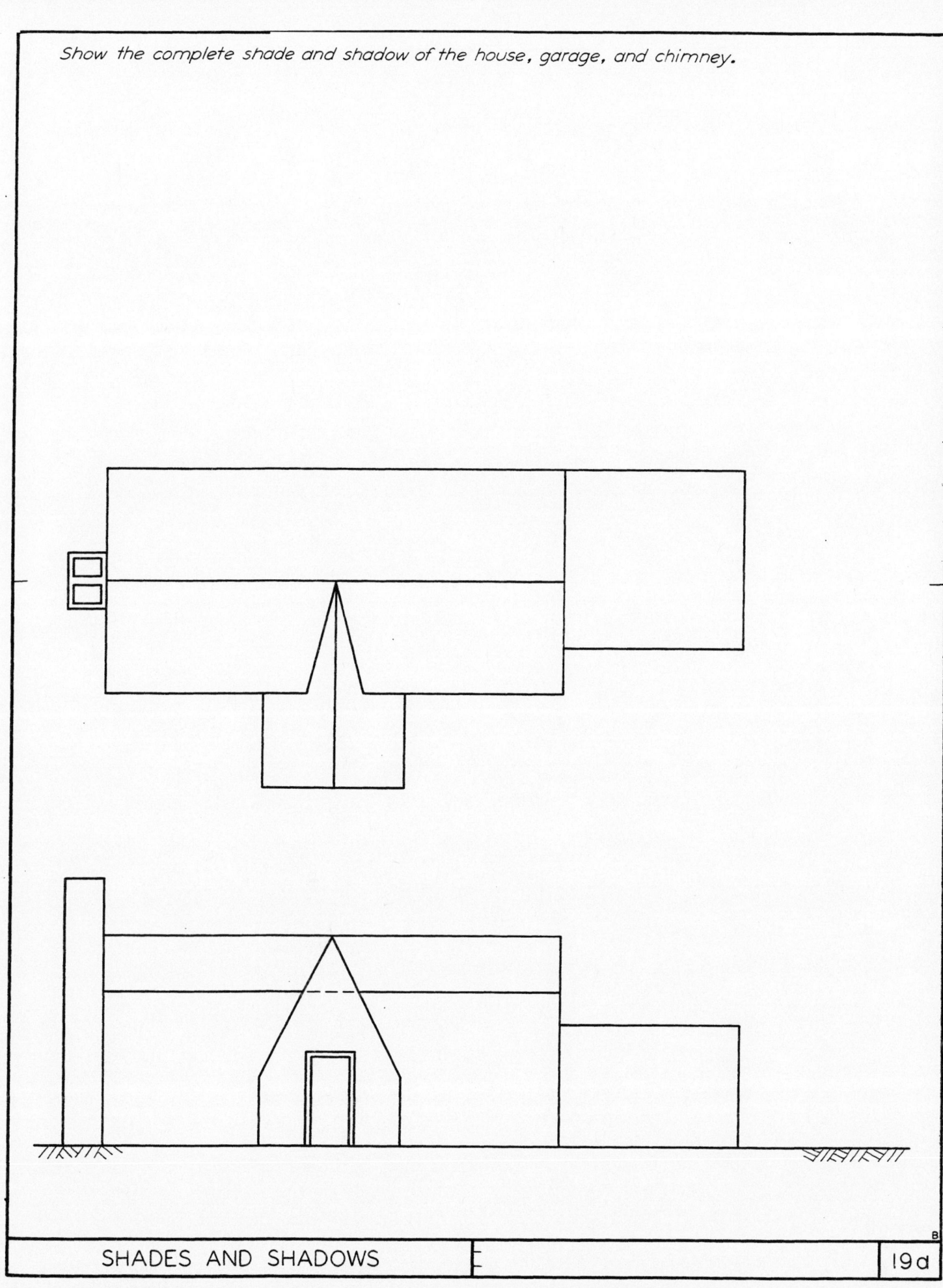

Show the complete shade and shadow of the house, garage, and chimney.

SHADES AND SHADOWS   19a

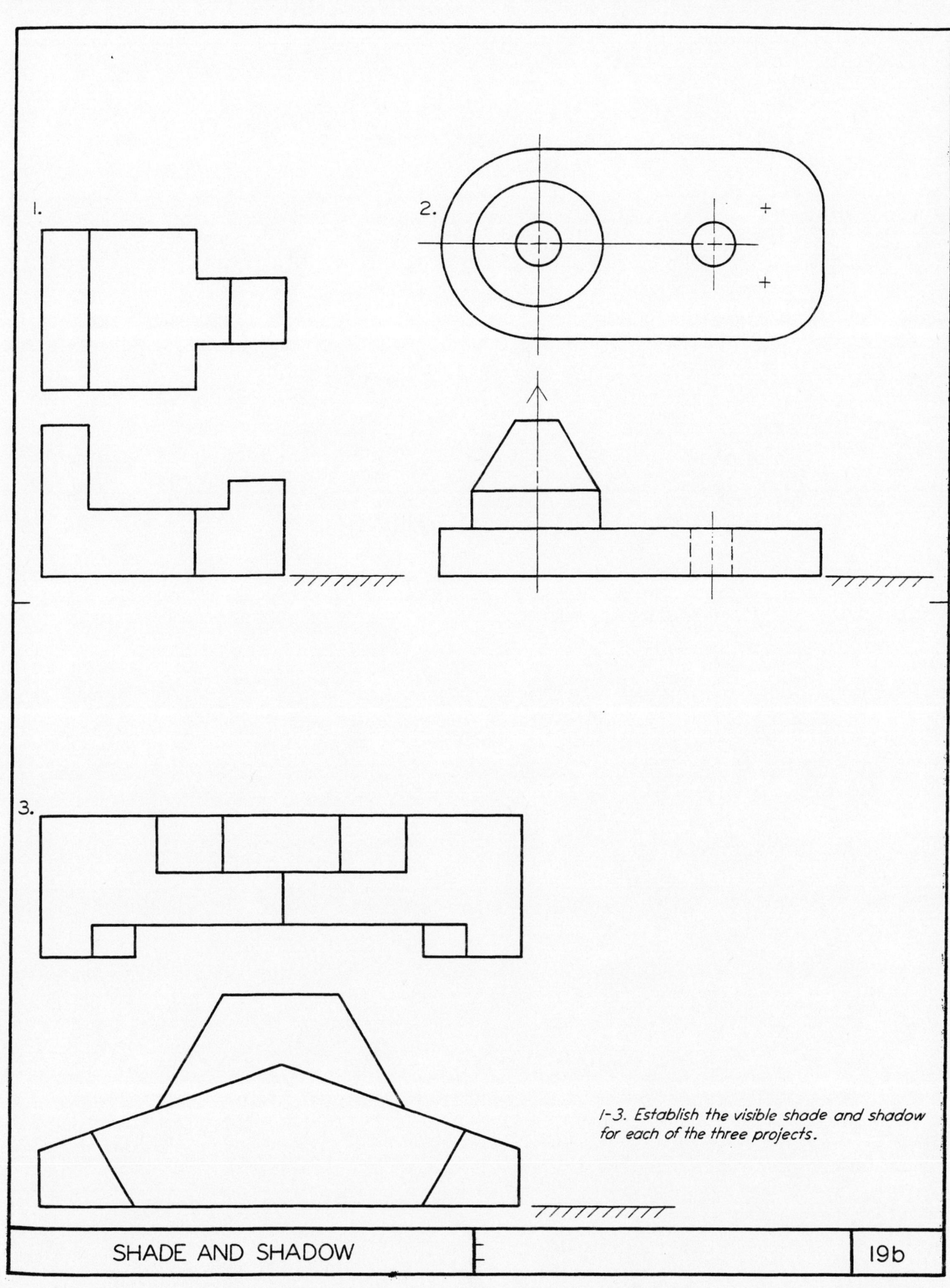

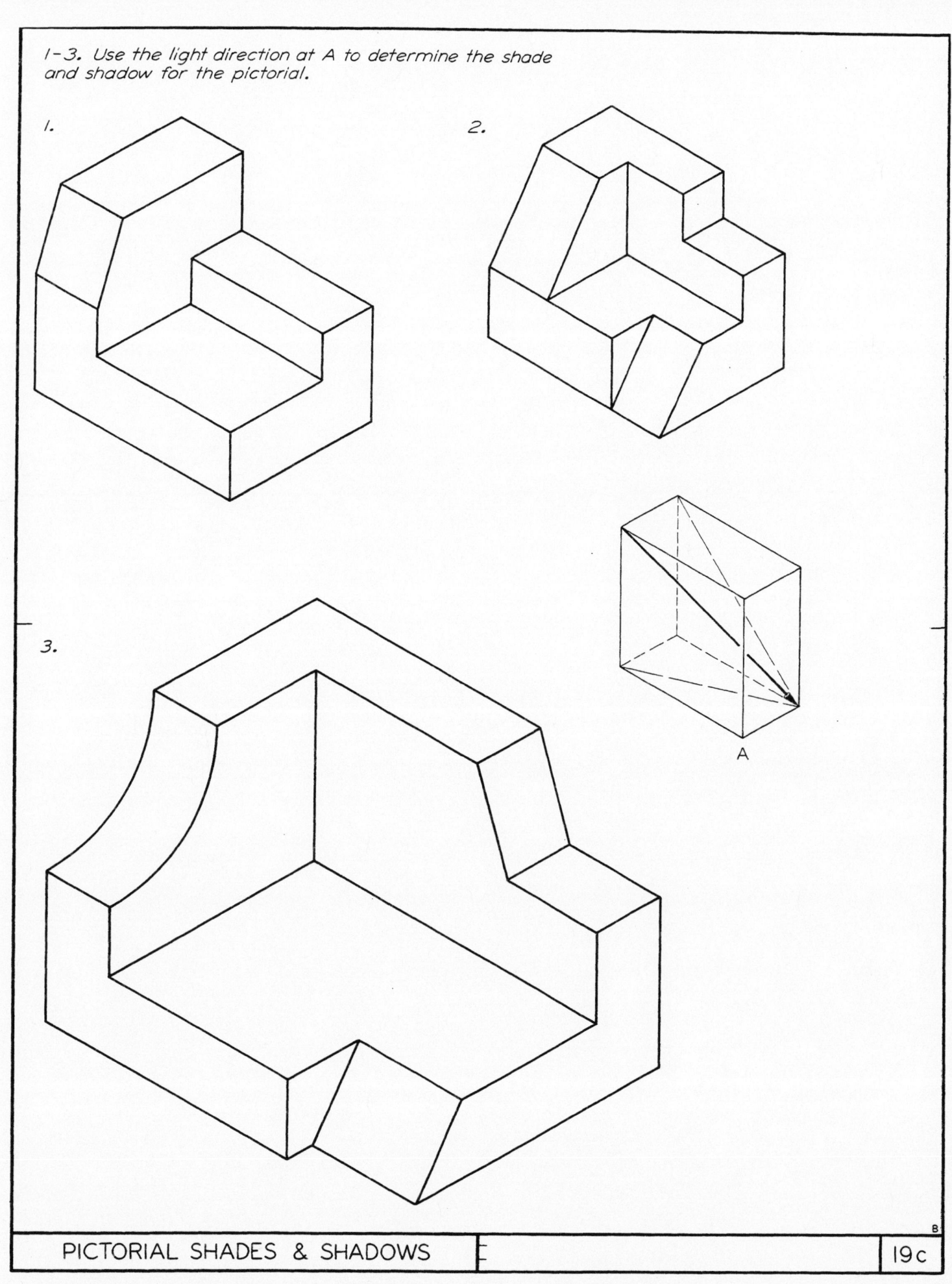

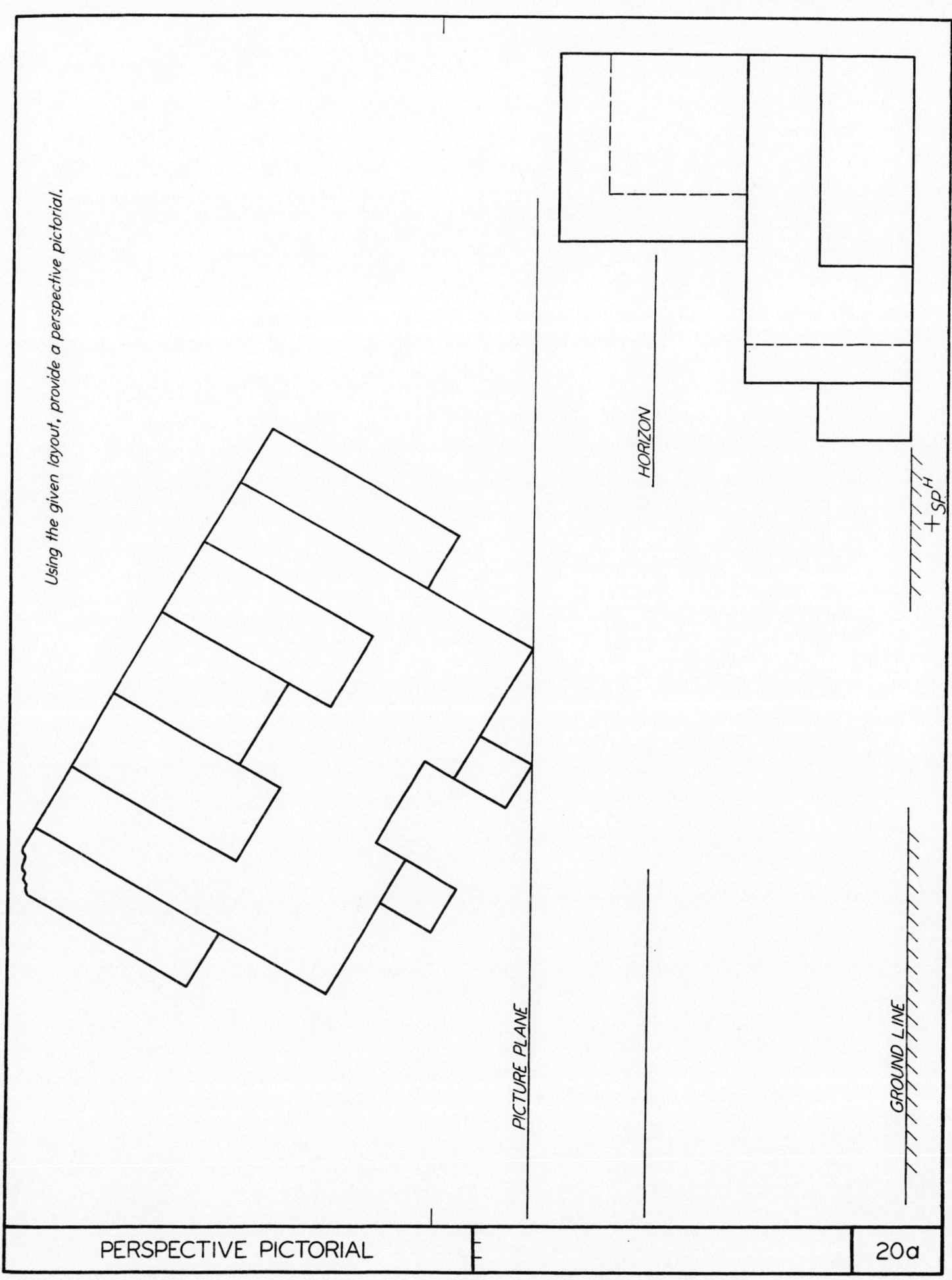

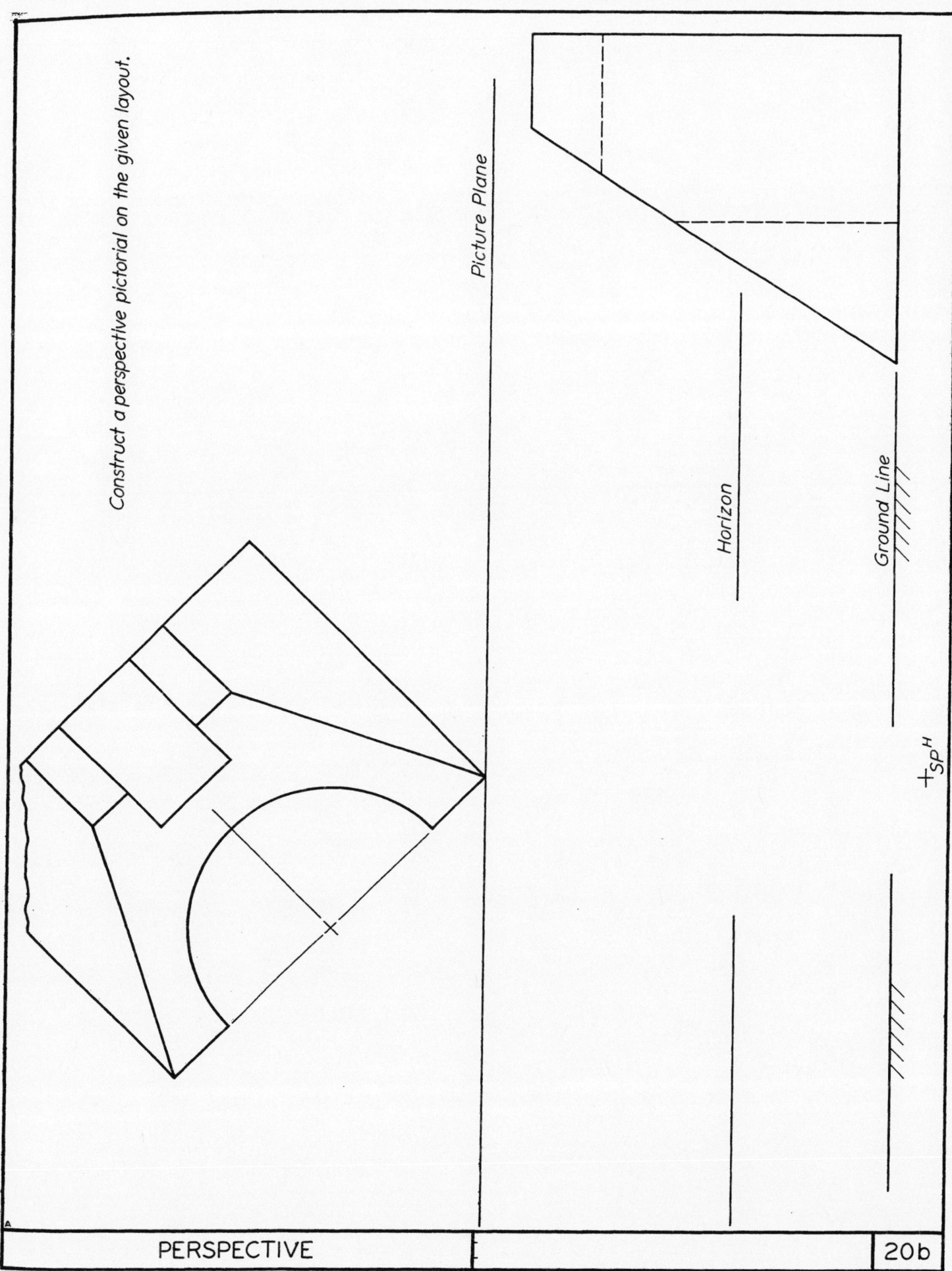

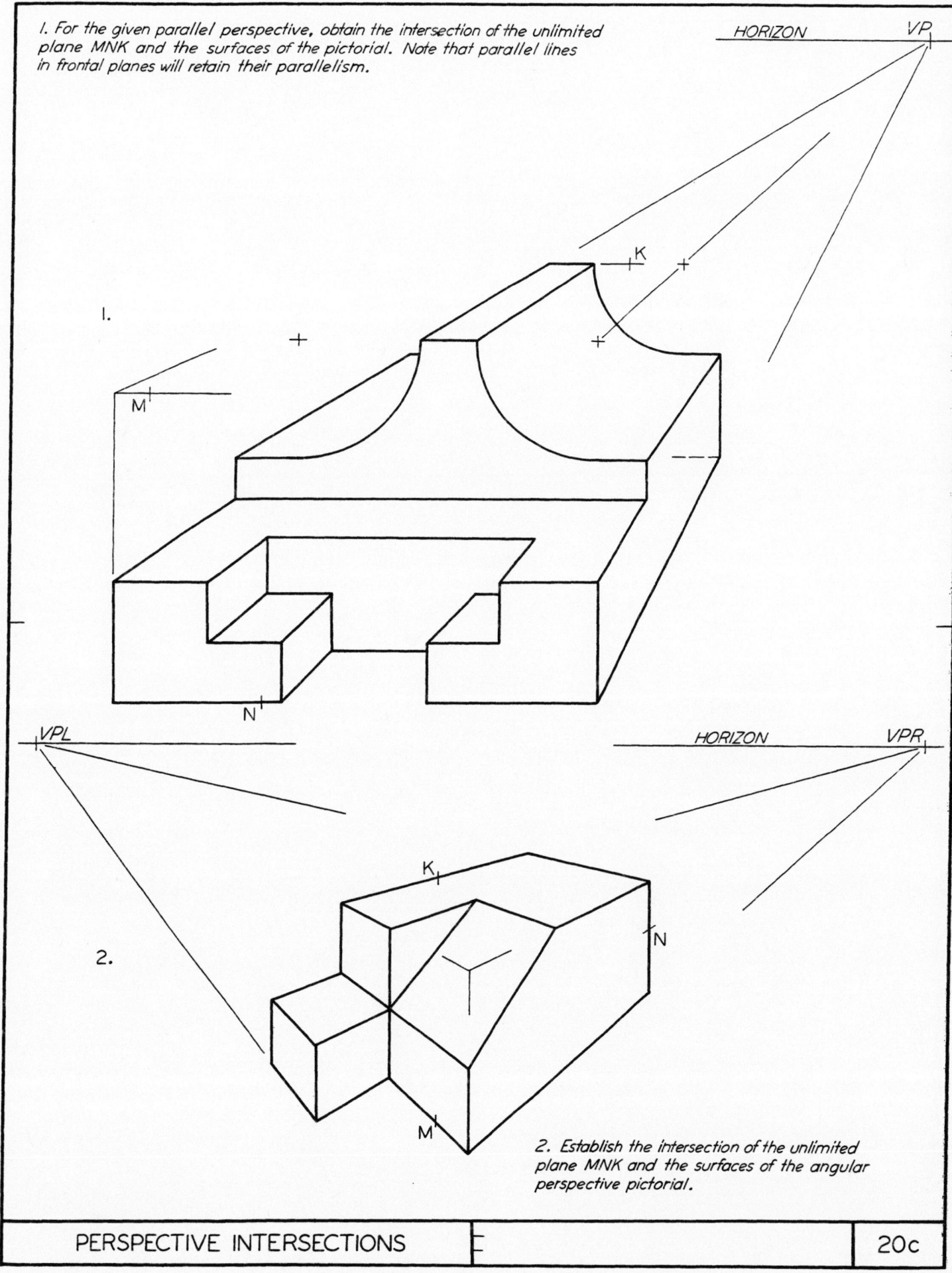

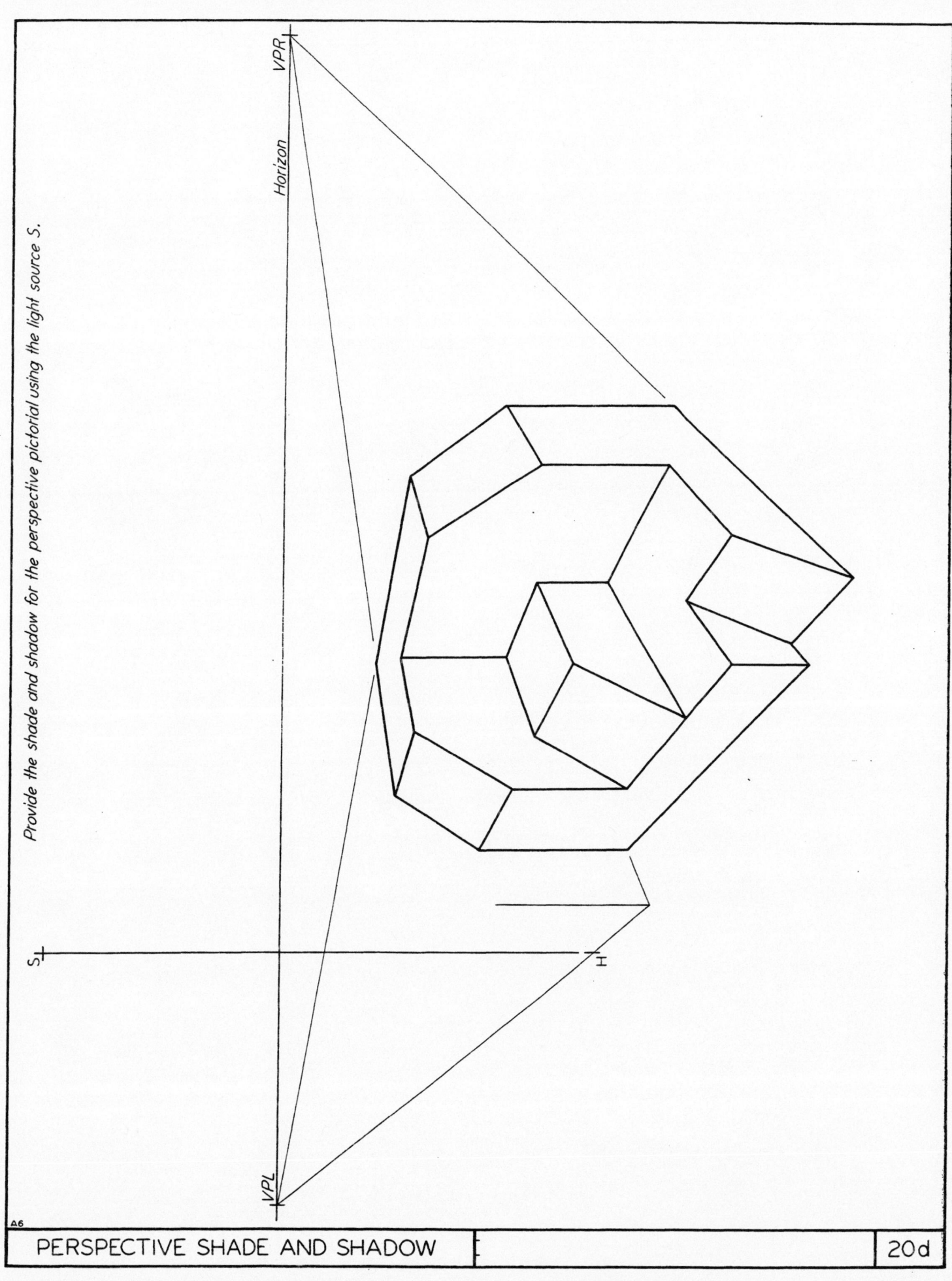

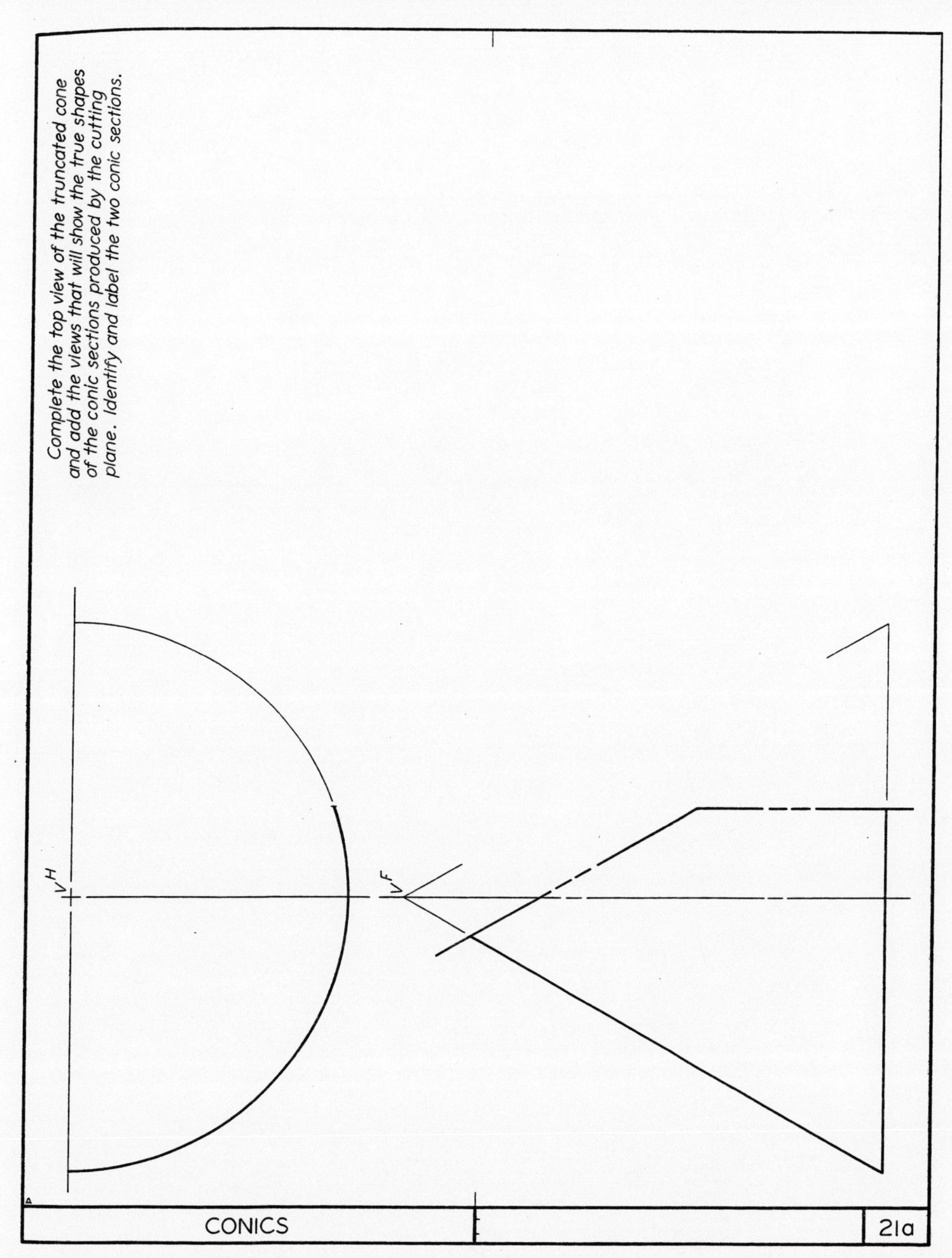

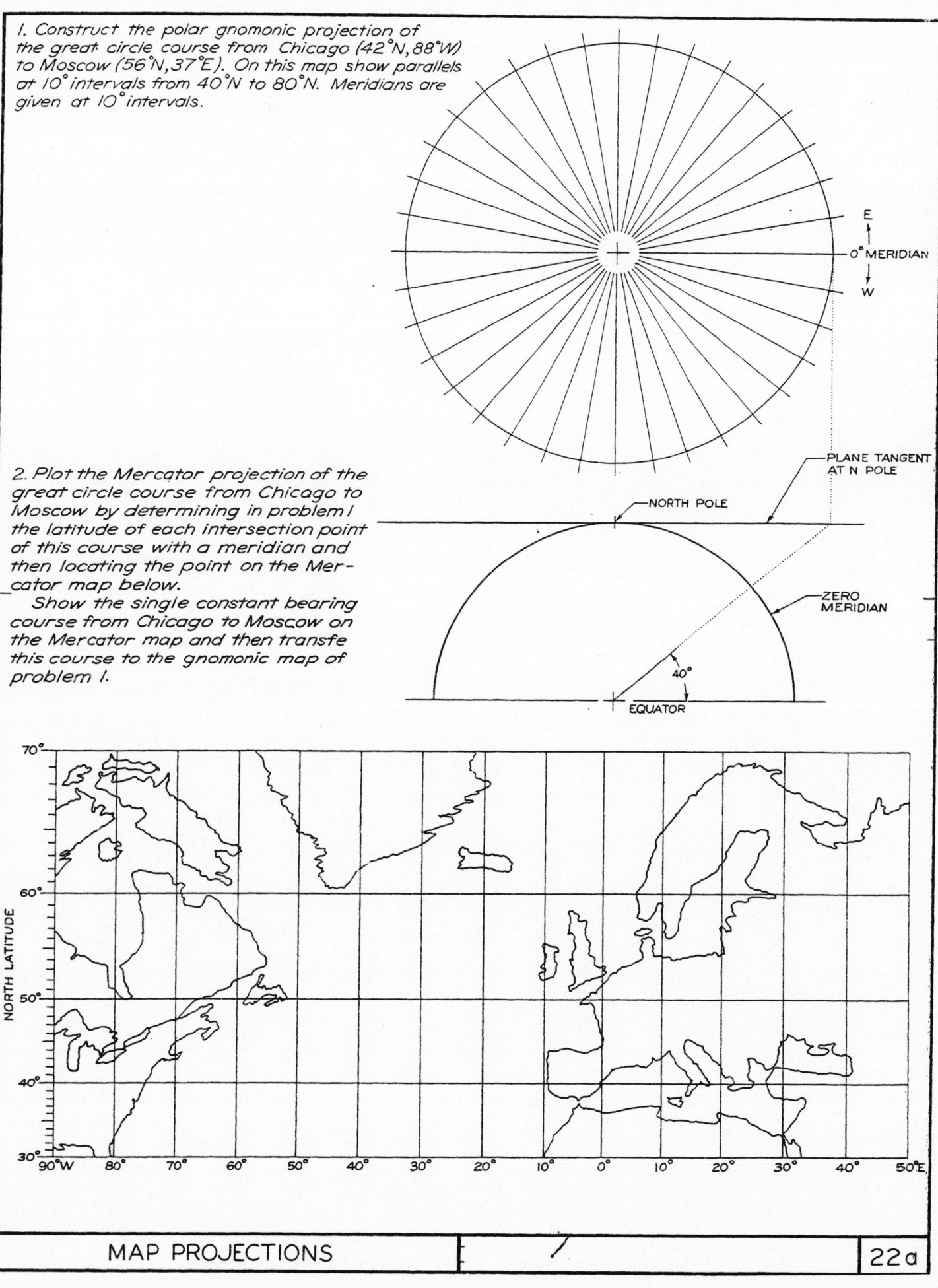

Locate the views of the spherical triangle ABC if side BC=45°, side CA=55°, and side BA=55°.
Find the angles of this triangle.

ANGLES-

A=..................
B=..................
C=..................

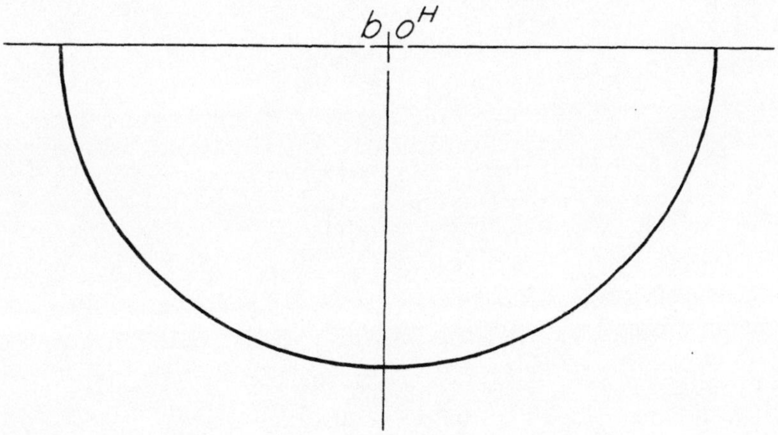

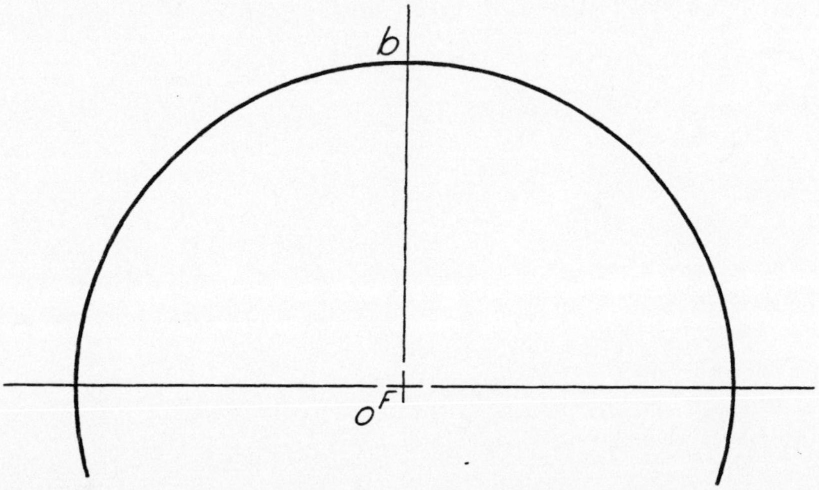

SPHERICAL TRIANGLE 23a

Determine the shortest distance along the earth's surface between point A, Lat. 45°N, Long. 145°W and point C, Lat. 40°N, Long. 10°W. (1° of arc ≅ 60 nautical miles.) Find the initial and final bearings of this great circle course from A to C.

Distance = _____
Initial Bearing = _____
Final Bearing = _____

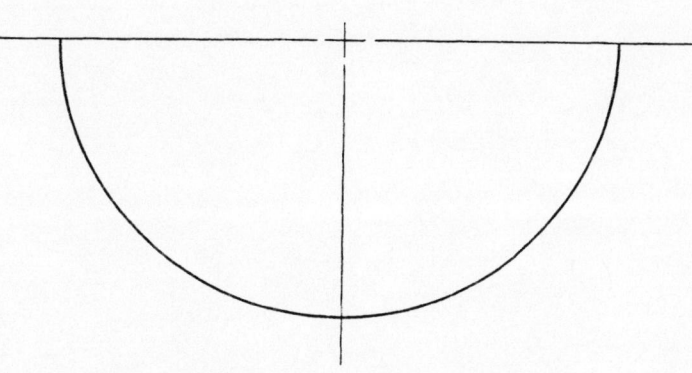

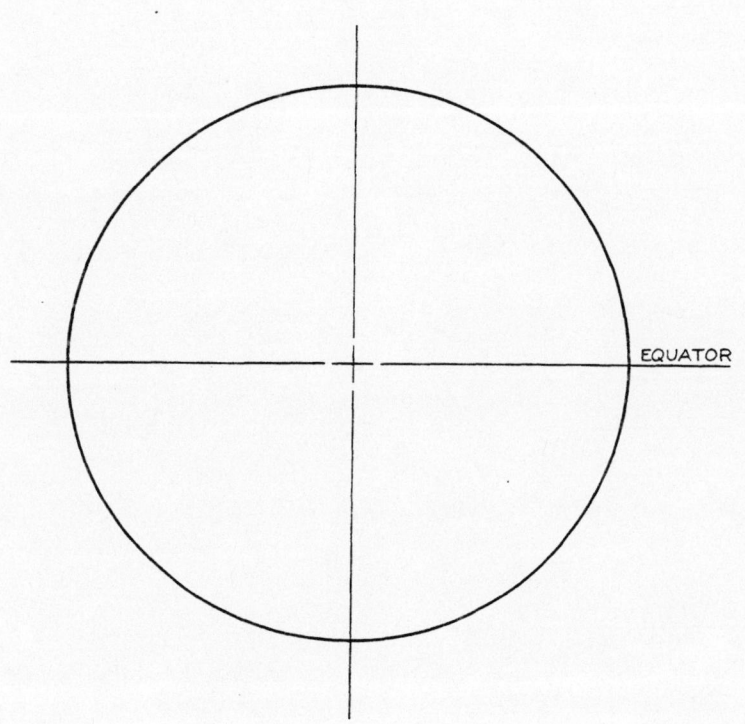

NAVIGATION

23 b

Determine the following:
  a. *The bearing and grade of line MN.*
  b. *The strike and dip of plane MNV.*
  c. *The true size of plane MNV.*
  d. *The dihedral angle formed by the planes MNV and MNO.*

Check (✓) the following for true or false: T   F
  1. *Lines OM and VN are parallel.*
  2. *Visibility of the given views is correct.*
  3. *Angle θ is the true slope of MN.*
  4. *View 4 will show a point view of line MO.*

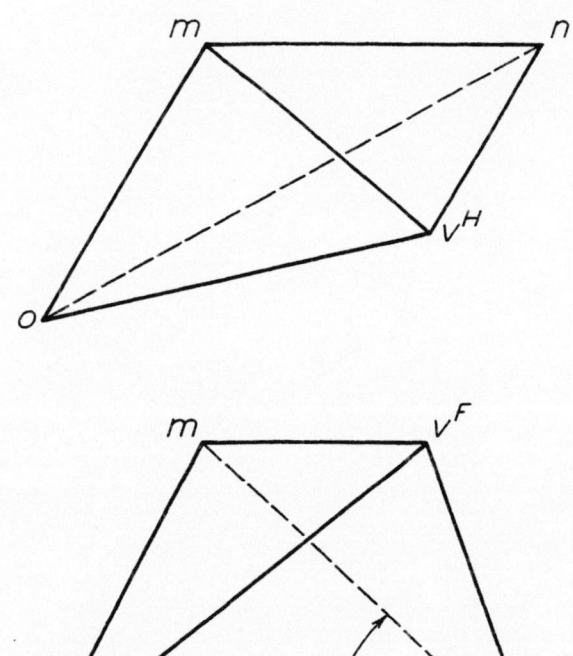

BASIC REVIEW   24a

Determine the following:
a. Bearing and grade of line AB.
b. True size of plane BCE.
c. True angle (θ_P) between line AC and a plane, P.
d. Intersection of line MN and the tetrahedron.

Check (✓) the following as true or false statements:

T	F	
		1. The visibility of the pyramid is correct.
		2. Lines BE and AC are parallel.
		3. Angle θ shows the true slope angle θ_H for line AC.
		4. View 4 will produce a point view of line MN.

BEARING -
GRADE -
θ_P -

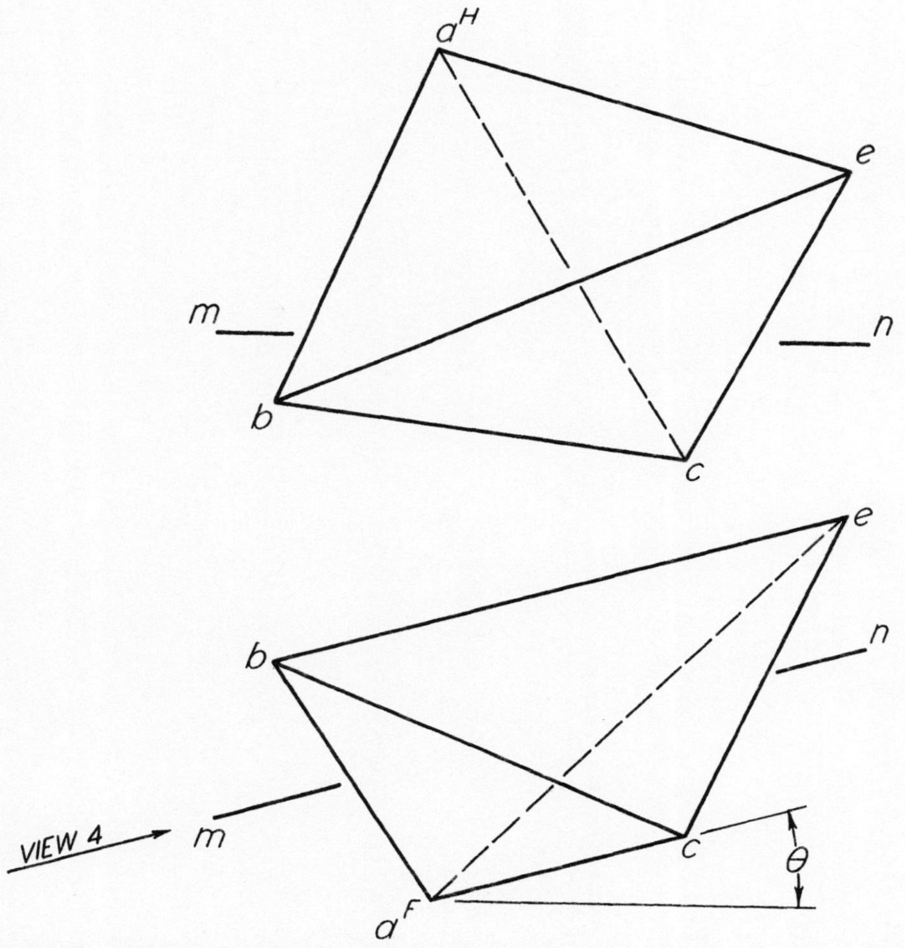

BASIC REVIEW PROJECT

24b

1. Obtain the intersection of line AB and the cylinder and line MN and the cone.

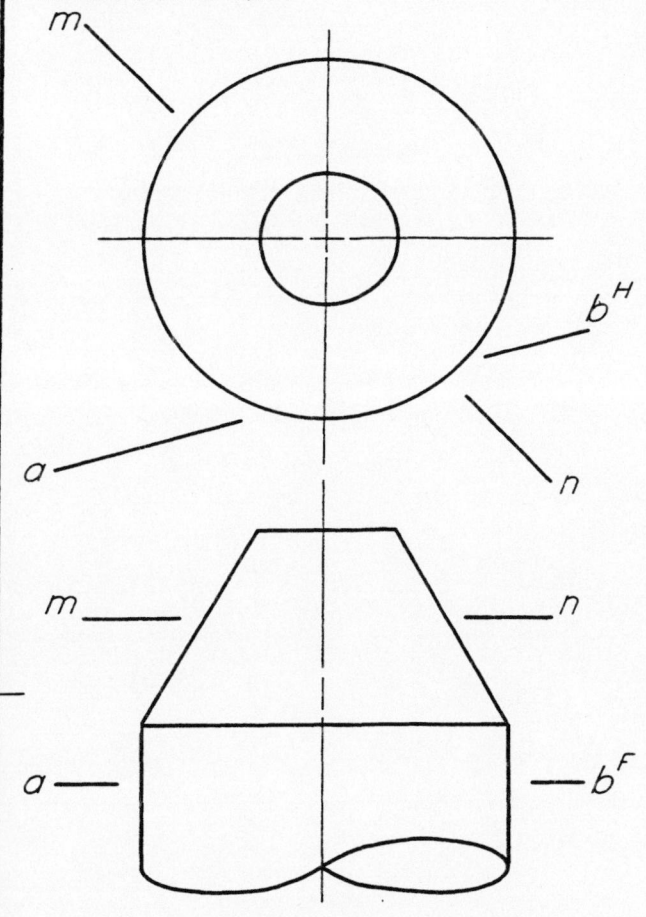

2. Locate the intersection of line AC and the cone and line NK and the cone.

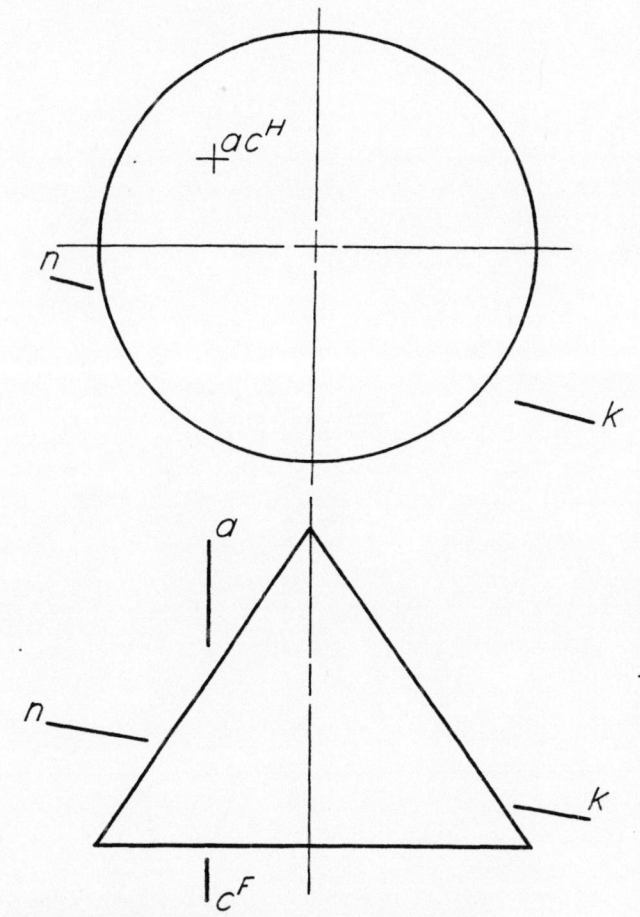

3. Locate the intersections of lines AE and MK with the spherical surface.

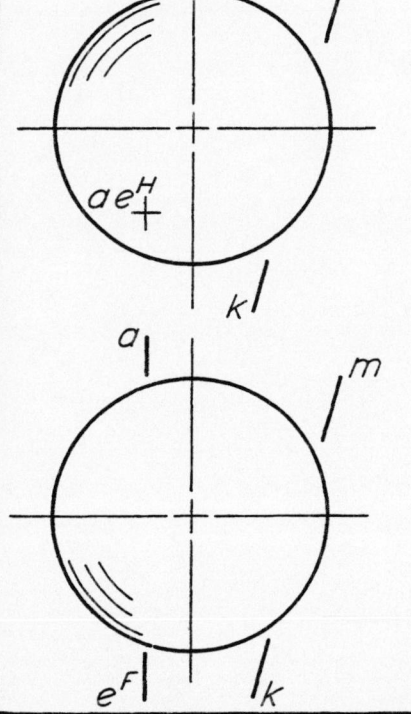

4. Establish the intersection of line CE and the cylindrical surface.

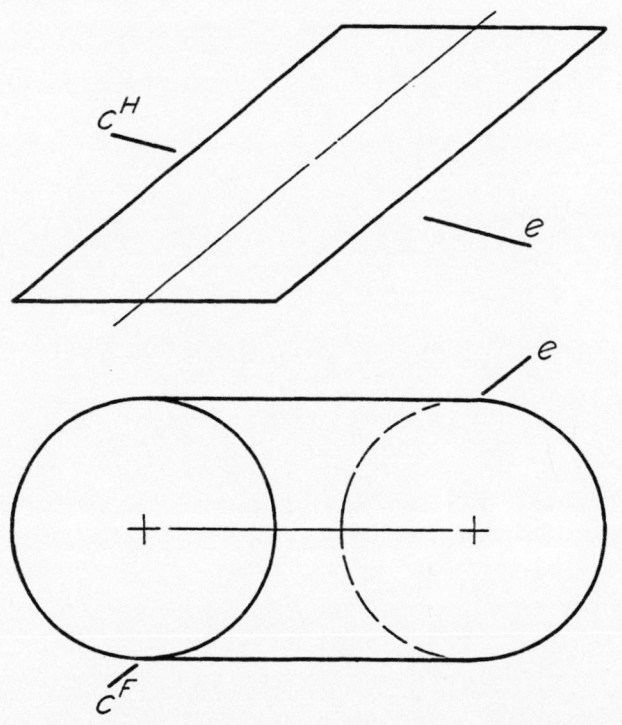

SPHERICAL INTERSECTIONS   24c

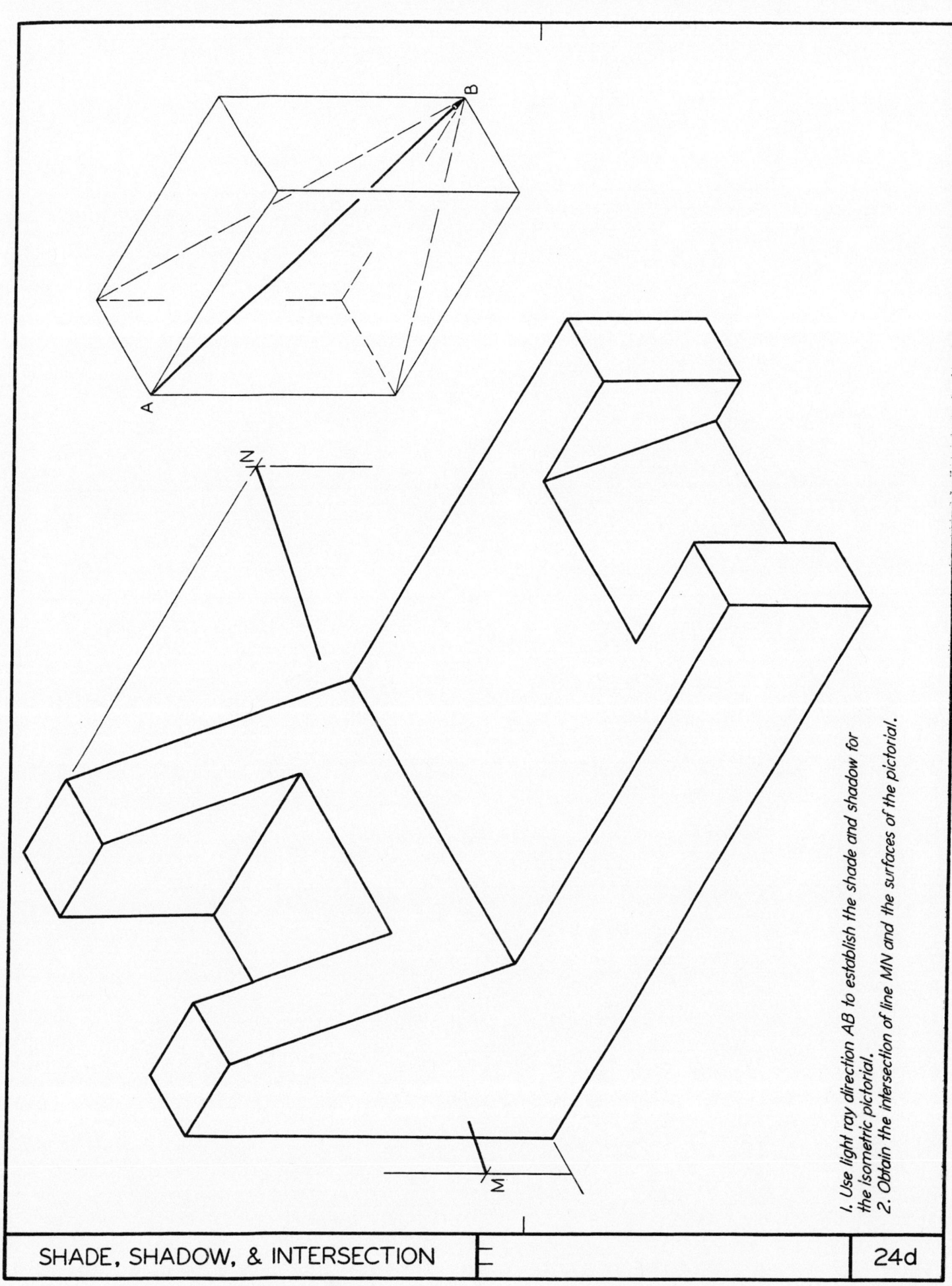

SHADE, SHADOW, & INTERSECTION 24d

1. Use light ray direction AB to establish the shade and shadow for the isometric pictorial.
2. Obtain the intersection of line MN and the surfaces of the pictorial.

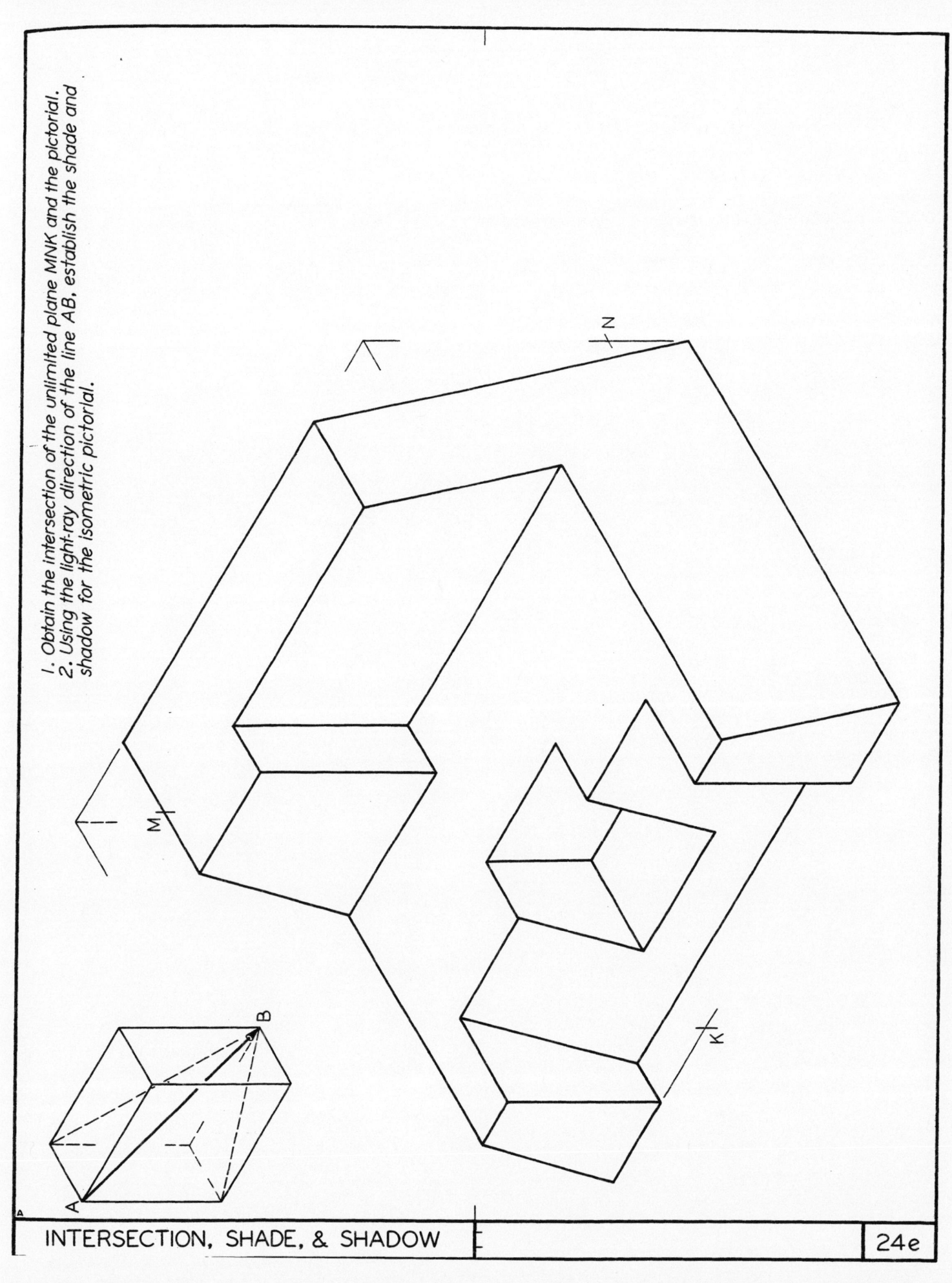

INTERSECTION, SHADE, & SHADOW   24e

1. Provide contour lines at 2m intervals.
2. On an overlay show contour lines, clubhouse, golf course, tennis courts, parking, etc.
3. List items for a cost estimate.
4. Design an attractive cover sheet for the report.

Scale: 1/4000.

LAKE EL 70.0

ROADWAY

GOLF COURSE LAYOUT

24f

1. Provide contour lines at 2 meter intervals.
2. On this original sheet, experiment with the desirable locations of a skateboard area and roller skate paths. Scale = 1:500
3. Show contour lines and facilities on an overlay.
4. If assigned, include additional facilities such as a water slide, picnic tables, clubhouse, trees, shrubs, etc.

73.6	73.4	74.0	75.2	74.2	74.6	75.0	75.2	73.4	73.6	73.8
73.2	75.4	75.4	76.2	77.4	77.6	76.6	75.8	75.4	75.2	74.0
75.2	76.2	77.0	78.0	78.2	78.4	77.0	77.6	77.2	77.4	77.0
75.0	77.4	79.0	81.4	78.6	77.2	77.2	79.4	81.2	78.8	77.4
77.0	78.2	81.0	77.4	70.6	73.6	75.2	79.4	77.6	81.4	79.4
77.0	81.6	79.0	75.4	77.4	81.4	80.6	75.0	80.8	78.2	77.2
77.0	78.4	78.2	81.0	81.2	78.6	77.8	80.6	77.0	77.2	75.6
75.8	77.2	78.4	78.8	79.0	76.6	75.6	77.0	75.2	74.8	75.0
75.4	76.0	77.0	77.2	76.6	75.8	75.0	73.8	73.6	73.0	72.8

ROADWAY

SKATEBOARD FACILITY

24g

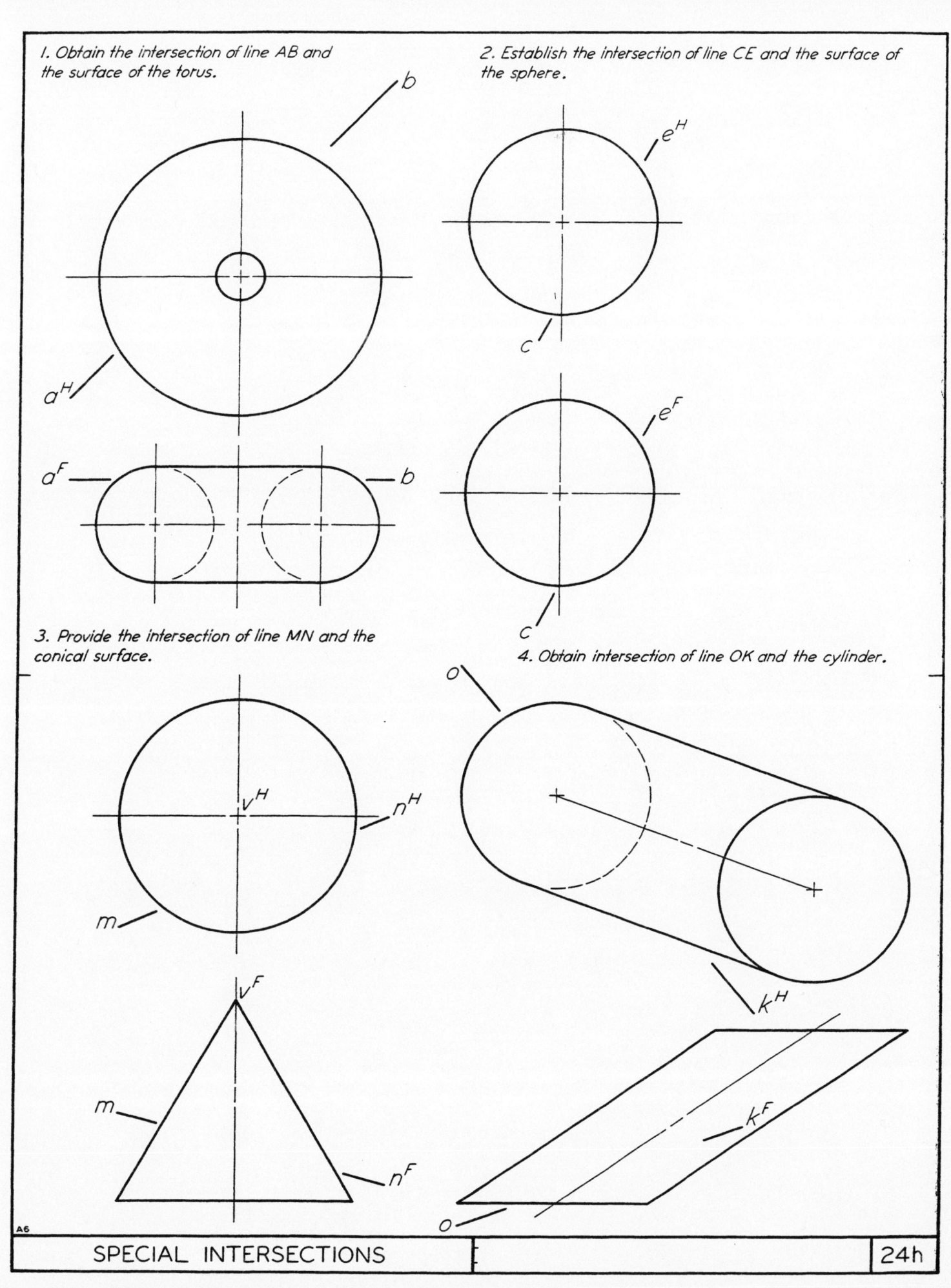

An aircraft at A is flying N-45° and diving at a rate of 2000 m per minute. The indicated air speed is 180 knots. A ship at S is proceeding N-300° at 42 knots. How close will the two pass? Show their locations at the moment of closest approach. Distance scale 1:100 000
Velocity scale 1mm=100m/min

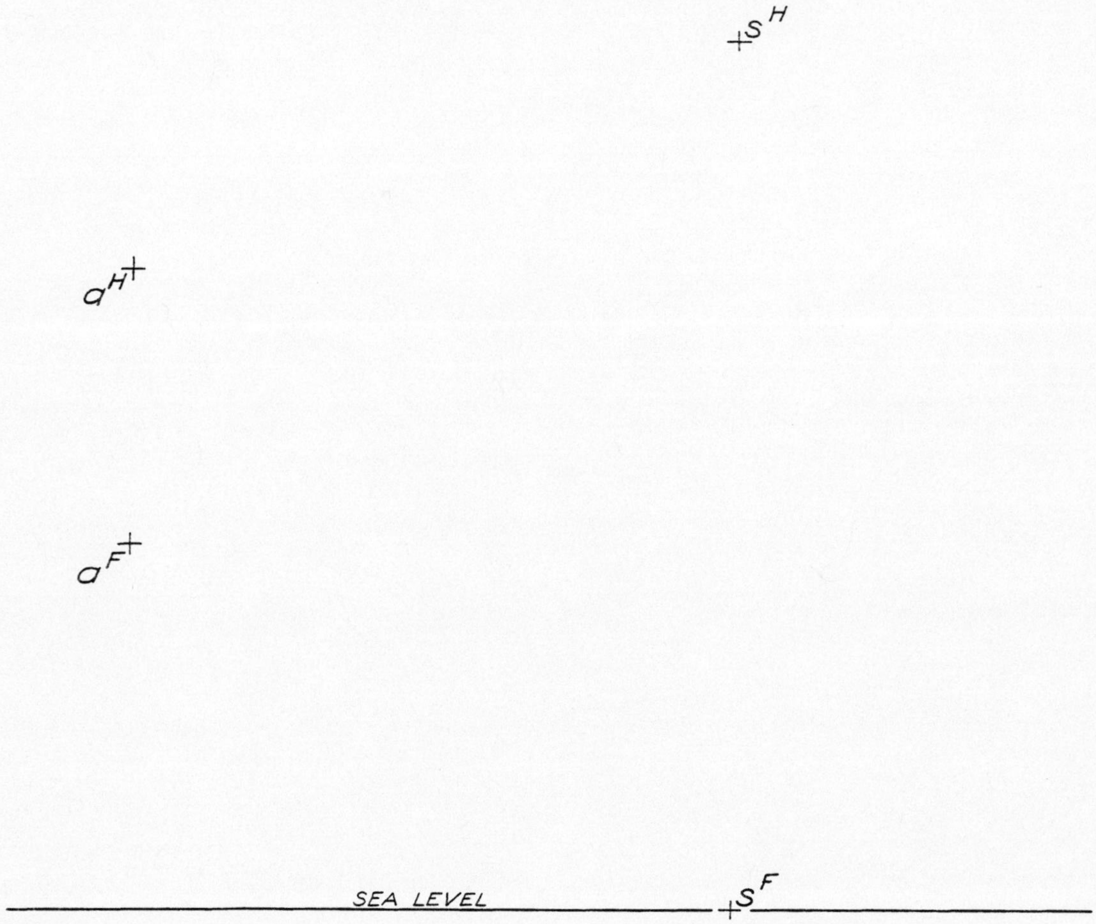

+S^H

a^H+

a^F+

_____ SEA LEVEL _____ +S^F _____

CALCULATIONS:

NON-COPLANAR VECTORS 24i

IN THIS INTRODUCTION TO COMPUTER GRAPHICS, DESCRIPTIVE GEOMETRY PROJECTS HAVE BEEN SELECTED THAT ENTAIL EITHER REPETITIVE CALCULATIONS OR MULTIPLE SOLUTIONS, FEATURES THAT SHOULD EXIST TO TAKE MAXIMUM ADVANTAGE OF THE COMPUTER CAPABILITIES. THE PROJECTS ARE INTRODUCED IN ALPHANUMERIC TERMINOLOGY, THAT IS IN GENERAL RATHER THAN NUMERICAL FORMAT. IN THE COMPUTER PROGRAMS, MATH CALCULATIONS ARE PRESENTED IN THESE GENERAL TERMS. THE OPERATOR WILL SUBSTITUTE ACTUAL NUMERICAL VALUES WHEN SO REQUESTED IN THE PRE-PREPARED PROGRAMS. AN ARRAY OF SUCH NUMERICAL VALUES ARE GIVEN THAT PERMIT A SUBSTANTIAL VARIETY OF SELECTIONS FOR CLASS USE. AT WASHINGTON STATE UNIVERSITY, WE CURRENTLY UTILIZE THE HEWLETT-PACKARD TERMINAL NO.2623A WITH VIDEO CATHODE RAY TUBE (CRT) DISPLAY OUTPUT. THE TERMINAL ESSENTIALLY PROVIDES A TYPEWRITER KEYBOARD WITH ADDITIONAL CHARACTERS AND CONTROLS. EACH OF THE SEVERAL TERMINALS IS WIRE CONNECTED TO A PRIME 2250 MINI-COMPUTER THAT PERFORMS THE OPERATIONS SPECIFIED IN THE COMPUTER PROGRAM. ALSO AVAILABLE IS A TEKTRONIX 4662 INTERACTIVE DIGITAL PLOTTER THAT PRODUCES HARD-COPY DRAWINGS EQUIVALENT TO THAT PREPARED BY A DRAFTER. THE COMPUTER PROGRAMS HAVE BEEN PRE-PREPARED EMPLOYING FORTRAN GRAPHIC SUPPORT LANGUAGE PLOT 10. THESE PROGRAMS MAY REQUIRE SOME MODIFICATION FOR THE EQUIPMENT AT OTHER SCHOOLS. TO ACTIVATE THE COMPUTER EQUIPMENT, AN ON OFF SWITCH MUST BE SET TO ITS 'ON' POSTION. THIS SWITCH ON THE HEWLETT-PACKARD TERMINAL IS LOCATED AT THE REAR OF THE VIDEO DISPLAY CONSOLE. ALSO IN THIS SAME LOCATION IS A LINE/LOCAL SWITCH OR CONNECTION THAT IS NORMALLY PRE-SET TO ENGAGE WITH THE PRIME COMPUTER RATHER THAN HAVE THE TERMINAL OPERATE AS AN INDEPENDENT TYPEWRITER. TO GAIN ACCESS TO THE PRIME COMPUTER AND THE STORED PROJECTS FOR THE ME 102 DESCRIPTIVE GEOMETRY COURSE, YOU MUST INITIALLY 'LOGIN'. THIS ESSENTIAL STEP IS ACCOMPLISHED BY TYPING THE COMMAND 'LOGIN' FOLLOWED BY THE ASSIGNED 'USER-ID' (USER-IDENTIFICATION). FOR ME 102 STUDENTS AT WSU, THE 'USER-ID' IS CONVENIENTLY THE COURSE NUMBER ME102. A UNIQUE PASSWORD MAY THEN BE REQUIRED TO HINDER UNAUTHORIZED ACQUISITION TO THE STORED DESCRIPTIVE GEOMETRY COMPUTER PROGRAMS CALLED THE 'USER' FILE DIRECTORY (UFD). THE PASSWORD IS ME102.

THIS INITIAL PRESENTATION PROVIDES THE GRAPHICAL ANALYSIS TO SECURE THE TRUE LENGTH OF A LINE AB. ASSUME THAT THE FRONT AND TOP VIEWS ARE ESTABLISHED AS SHOWN IN THE GIVEN ILLUSTRATION. THE TRUE LENGTH OF LINE AB IS THE HYPOTENUSE OF A RIGHT TRIANGLE THE RISE OF WHICH IS EQUIVALENT TO DISTANCE Y. THE RUN OBSERVED IN TOP VIEW IS THE HYPOTENUSE OF A RIGHT TRIANGLE INVOLVING THE LEGS X AND Z.

SUGGESTED ARRAY OF VALUES FOR X, Y, AND Z

X	Y	Z
3	2	4
4	3	4.4
5	4	5.2
6	5	6
7	6	8
8	8	-2
9.4	8.4	0
-2	0	-2.4
0	-1.6	-2.8

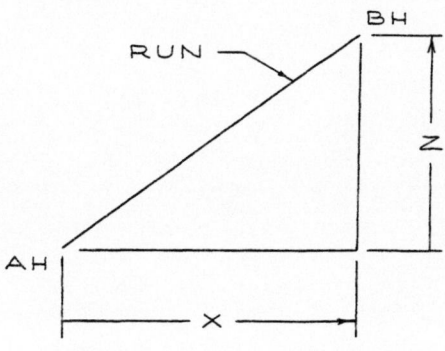

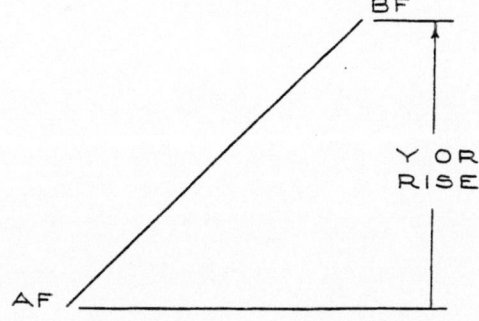

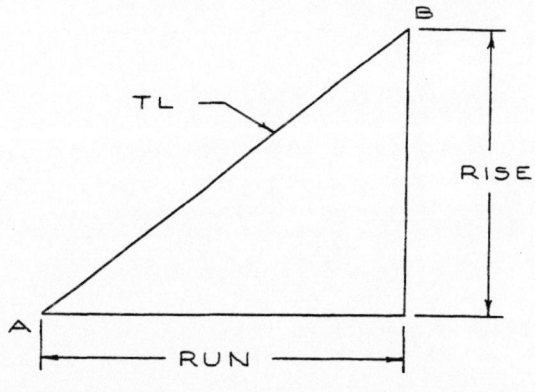

RELATED COMPUTER INPUT

GENERAL COMMENTS

COMMENTS ARE PROVIDED TO EXPLAIN THE PREPARATION OR USE OF THE COMPUTER PROGRAM.
ACTIVATE COMPUTER, TYPE LOGIN.
TYPE USER-ID, ME102

TYPE THE PASSWORD, ME102.
AFTER OK, SIGNAL APPEARS ON THE TERMINAL, TYPE LD. ME102 MENU WILL BE LISTED. SELECT TL OF LINE PROGRAM BY TYPING, TLLINE.
VARIABLES OUT(1) THRU OUT(10) WILL BE USED TO STORE DATA OF UP TO 80 CHARACTERS EACH.
CHAR *1 WILL BE USED TO STORE ONE-LETTER RESPONSES TO QUESTIONS. FOR EXAMPLE: Y FOR YES.
INTEGER *2 ID, I WILL BE USED TO STORE WHOLE-NUMBER VARIABLES.

THESE PROGRAM VARIABLES ARE IN FLOATING POINT MODE TO ACCEPT DECIMAL VALUES.
THIS ENTRY IS USED TO INDICATE WHERE TO LOCATE DATA OR ANS. FOR EXAMPLE: 0 FOR THE TERMINAL, 22 FOR THE DIGITAL PLOTTER.

```
LOGIN
USER-ID
PASSWORD
OK,
TLLINE

CHARACTER *80 OUT(10)

CHARACTER *1 ANS

INTEGER *2 ID, I

REAL *4 X,Y,Z, RISE, RUN, TLAB, GRADE, SLOPE

ID = 0
```

COMPUTER GRAPHICS

1. STUDY THE ANALYSIS AND MATH THAT PRODUCE THE INTERSECTION OF LINE MN WITH PLANE ABCE IN THE OBLIQUE STYLE PICTORIAL. PREPARE A COMPUTER PROGRAM TO DRAW THIS PICTORIAL IN WHICH LOCATION NY IS A SELECTED VARIABLE.

ARRAY OF NY VALUES (CM)

3.0	3.5	4.0
4.5	5.0	5.5

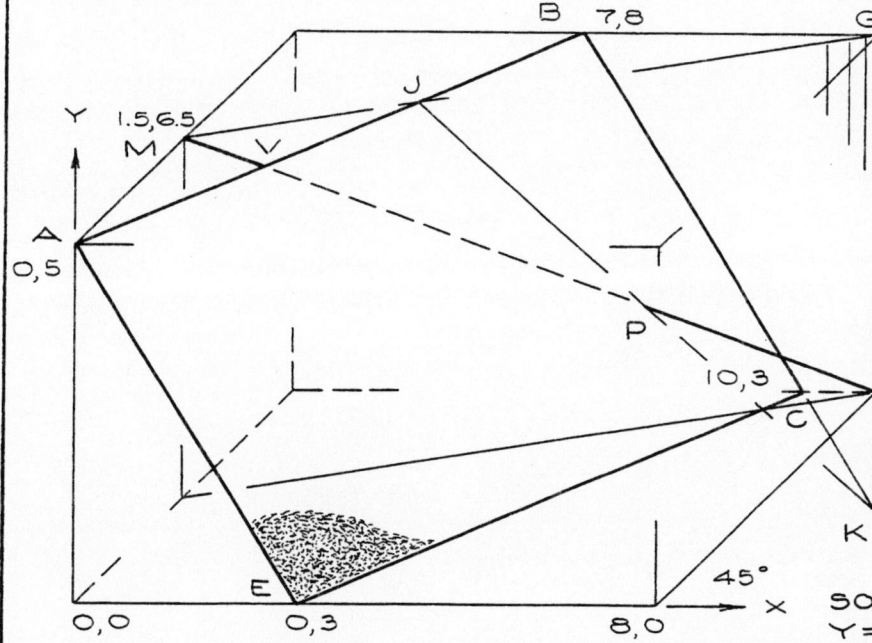

EQUATIONS AND MATH
LINE AB $Y = .429X + 5$
BC $Y = -1.667X + 19.667$
LINE MN
$Y = (-.684 + .1053 NY)X + 7.526 - .158 NY$
LINE MG $Y = .158X + 6.263$
LINE GN $X = 11$
SOLVE FOR LOCATION J :
$.429X + 5 = .158X + 6.263$ OR
$JX = 4.66$, $JY = .429(4.66) + 5$
OR $JY = 7.0$

SOLVE FOR KY ON THE LINE BC WHEN $KX = 11$:
$Y = -1.667(11) + 19.667$ OR
$KY = 1.33$
LINE JK EQUATION :
$Y = -.894X + 11.17$
SOLVE FOR LOCATION P :
$Y = (-.684 + .1053 NY)X + 7.526 - .158 NY = -.894X + 11.17$ OR
$PX = \dfrac{3.644 + .158 NY}{.21 + .1053 NY}$ AND
$PY = -.894(PX) + 11.17$
SOLVE FOR LOCATION V :
$.429X + 5 = (-.684 + .1053 NY)X + 7.526 - .158 NY$ OR
$VX = \dfrac{2.526 - .158 NY}{1.113 - .1053 NY}$ AND
$VY = .429 VX + 5$

2. IN THIS OBLIQUE PICTORIAL, OBTAIN GRAPHICALLY THE INTERSECTION OF LINE KN WITH THE PLANE ABC. IF ASSIGNED, PREPARE THE MATH AND A COMPUTER PROGRAM TO DRAW THIS PICTORIAL.

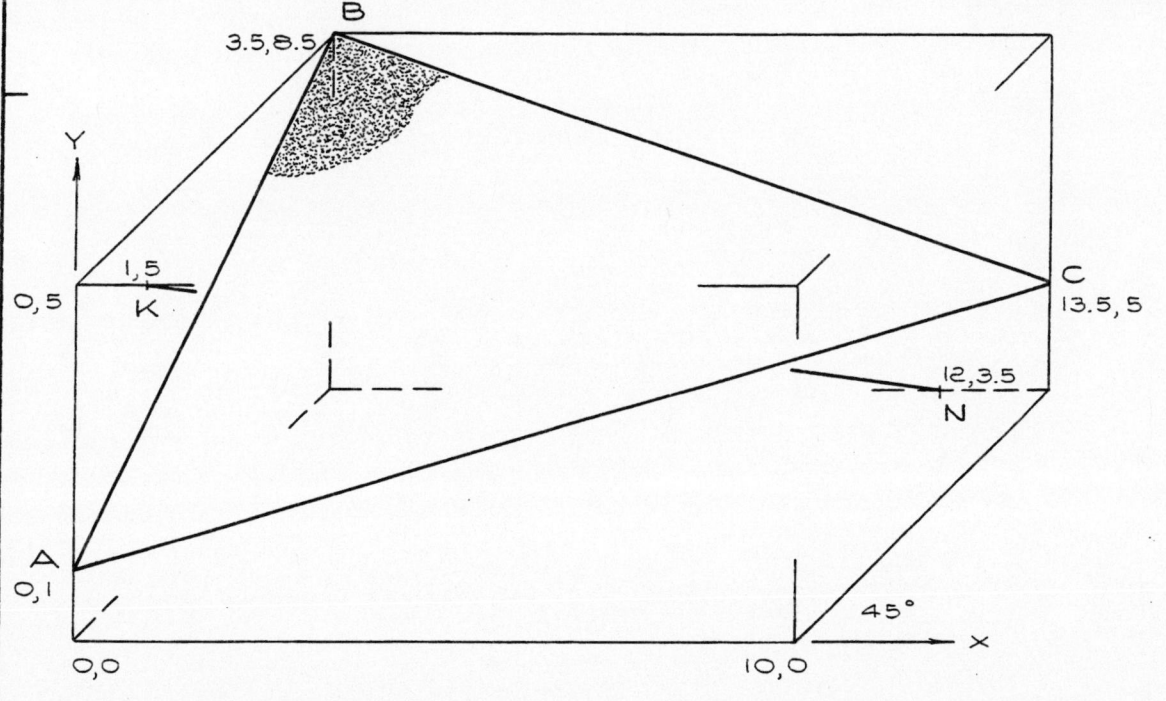

COMPUTER GRAPHICS   25 b

FOR THE OBLIQUE PICTORIAL, PROVIDE A COMPUTER
PROGRAM THAT ESTABLISHES THE INTERSECTION OF
LINE MN AND PLANE ABCD. OBSERVE THE VERTICAL
CUTTING PLANE MRNQ USED TO LOCATE PIERCING
POINT P.
LOCATION MX ON LINE KB IS A LISTED VARIABLE
TO ACCOMMODATE A VARIETY OF SOLUTIONS.

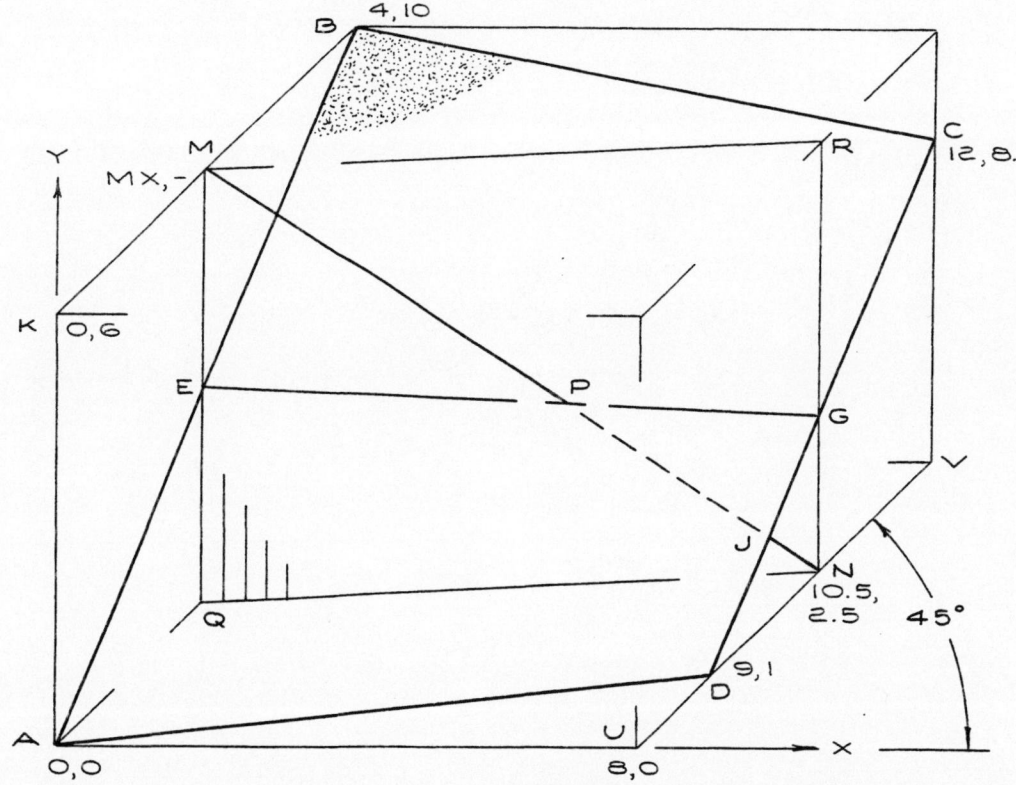

ARRAY OF MX VALUES

2	1.9
2.1	1.8
2.2	1.7
2.3	1.6
2.4	1.5
2.5	1.4
2.6	1.3
2.7	1.2
2.8	1.1
2.9	1.0
3	.9

MATH ANALYSIS

LINE AB EQUATION $Y = 2.5X$
LINE DC $Y = 2.5X - 21.5$
LINE MN $\dfrac{Y-Y_1}{X-X_1} = \dfrac{Y_1-Y_2}{X_1-X_2}$
NOTE THAT $MY = 6 + MX$
THEN $\dfrac{Y-(6+MX)}{X-MX} = \dfrac{6+MX-2.5}{MX-10.5}$
OR LINE MN EQUATION:
$$Y = \dfrac{(3.5+MX)X - 3.5MX - \overline{MX}^2}{MX-10.5} + 6 + MX$$

SOLVE FOR EY WHEN $EX = MX$
OR $EY = 2.5 MX$

SOLVE FOR GY WHEN $GX = 10.5$
$GY = 2.5(10.5) - 21.5$ THEN
$GY = 4.75$

OBTAIN LINE EG EQUATION:
$\dfrac{Y - 2.5MX}{X - MX} = \dfrac{2.5MX - 4.75}{MX - 10.5}$ OR

$$Y = \dfrac{(2.5MX - 4.75)X + 4.75MX - 2.5\overline{MX}^2}{MX - 10.5} + 2.5MX$$

SOLVE FOR PX, PY:
$3.5X + MX(X) - 8MX - 63 =$
$2.5 MX(X) - 4.75X - 21.5 MX$ OR

$PX = \dfrac{13.5 MX + 63}{8.25 + MX}$ AND

$PY = \dfrac{(2.5 MX - 4.75) PX - 21.5 MX}{MX - 10.5}$

SOLVE FOR JX, JY:
$\dfrac{3.5X + MX(X) - 8MX - 63}{MX - 10.5} = 2.5X - 21.5$

THEN $JX = \dfrac{288.75 - 13.5 MX}{29.75 - 1.5 MX}$

AND $JY = 2.5 JX - 21.5$

FOR A SIMILAR PROJECT, FIX
LOCATION MX AT 2.0 AND
INTRODUCE POSITION N ON
LINE UV AS A VARIABLE.

COMPUTER GRAPHICS DPINTA 25c

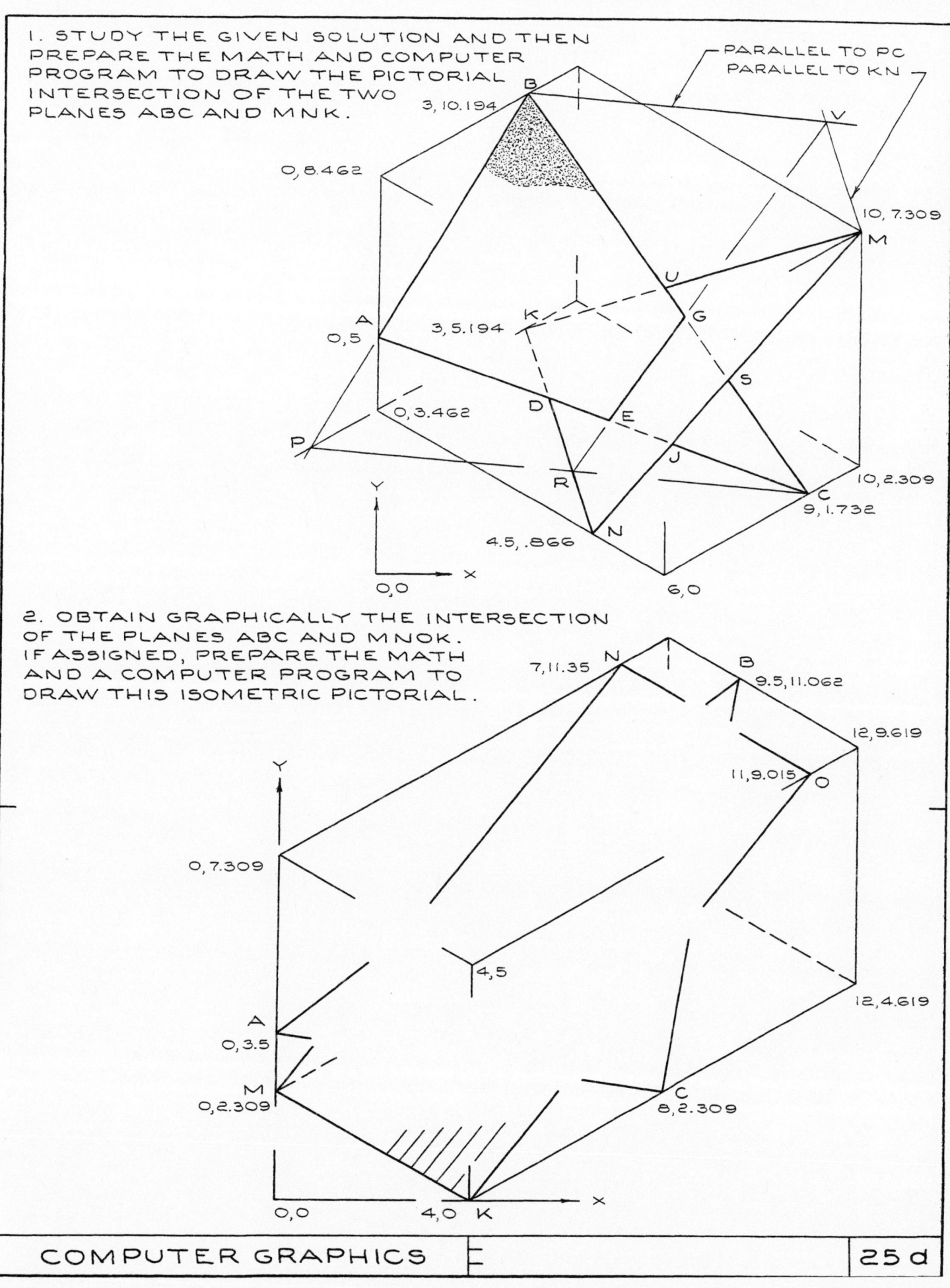

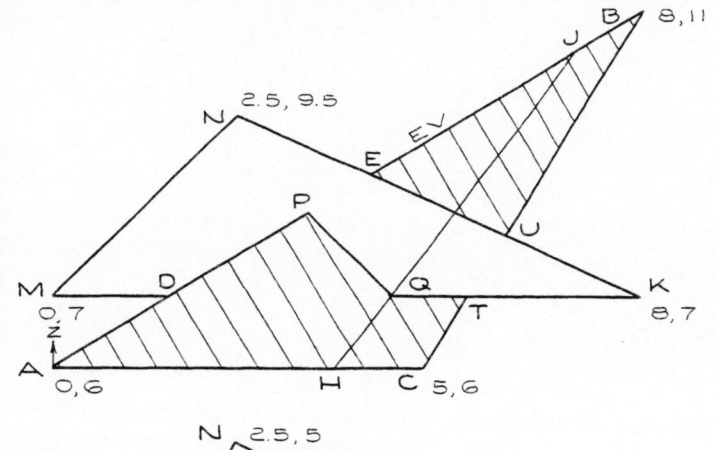

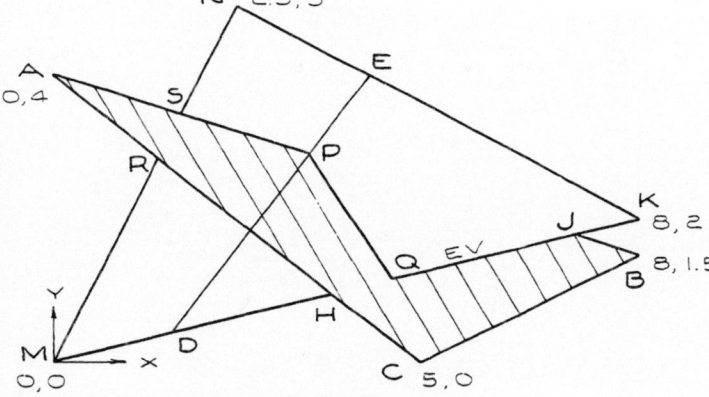

THIS PROGRAM PROVIDES THE INTER-
SECTION OF THE TWO PLANES ABC
AND MNK. ESSENTIAL PIERCING
POINT P IS OBTAINED USING AN EV
CUT THRU AB IN THE TOP VIEW.
PIERCING POINT Q IS SECURED
USING AN EV CUT THRU LINE MK
IN THE FRONT VIEW.

EQUATIONS OF PERTINENT LINES

TOP VIEW

AB	$Z = .625X + 6$
CB	$Z = 1.667X - 2.333$
NK	$Z = -.455X + 10.636$
MK	$Z = 7$
AC	$Z = 6$

FRONT VIEW

AB	$Y = -.3125X + 4$
AC	$Y = -.8X + 4$
MN	$Y = 2X$
MK	$Y = .25X$
NK	$Y = -.5454X + 6.364$

NOTE: INTRODUCE MECHANICAL OR
LETRATONE SHADING TO ENHANCE
THE VISIBILITY.

COMMENTS

ACTIVATE COMPUTER, TYPE LOGIN
USER-ID, TYPE ME102
PASSWORD, TYPE ME102
WHEN OK, APPEARS, TYPE LD
FROM MENU LIST, SELECT AND
TYPE THE PROGRAM, INPLA

BRIEF PROGRAM EXPLANATION.

ENTER SCALE FACTOR (SF)
FOR EXAMPLE: TYPE 1., 1.2, 1.5, OR .5
SOLVE FOR DX, DZ, DY: DZ = 7
$DX = \frac{DZ-6}{.625}$, DY = .25 DX
DX = 1.6, DY = 4
SOLVE FOR EX, EZ, EY:
$.625X + 6 = -.455X + 10.636$, $EX = \frac{4.636}{1.08}$
EZ = .625 EX + 6
EY = -.5454 EX + 6.364
EX = 4.293, EZ = 8.683, EY = 4.023
DE EQUATION (FRONT VIEW):
Y = 1.345X - 1.752,
SOLVE FOR PX, PY, PZ:
-.3125X + 4 = 1.345X - 1.752
PY = -.3125 PX + 4, PZ = .625 PX + 6
SOLVE FOR HX, HY, HZ:
.25X = -.8X + 4, $HX = \frac{4}{1.05}$
HY = .25 HX, HZ = 6

COMPUTER INPUT

```
LOGIN
USER ID
PASSWORD
OK,

INPLA
CHARACTER *1 ANS
INTEGER *2 ID, I
REAL *4 DX, DZ, DY, EX, EZ, EY, HX,
& HZ, HY, JX, JY, JZ, TX, TZ, UX, UZ, RX,
RY, SX, SY, PX, PY, PZ, QX, QY, QZ
ID = 0
PRINT *, 'PROGRAM INPLA'
PRINT *, 'USER PROVIDES SCALE'
PRINT *, 'THE PROGRAM DISPLAYS'
PRINT *, 'PLOT OF VIEWS'
100 CONTINUE
PRINT *, 'ENTER SCALE FACTOR'
READ(1, *, ERR = 100) SF
DZ = 7.
DX = (DZ - 6.)/.625
DY = .25 * DX
EX = 4.636/1.08
EZ = (.625 * EX) + 6.
EY = (-.5454 * EX) + 6.364



PX = 5.752/1.6575
PY = (-.3125 * PX) + 4.
PZ = (.625 * PX) + 6.
HX = 4./1.05
HY = .25 * HX
HZ = 6.
```

COMPUTER GRAPHICS (INTERSECTING PLANES)

REVIEW SHEET CG1 FOR LOGIN INFORMATION. THIS PROJECT INVOLVES THE SHEET-METAL DEVELOPMENT OF A CONE FRUSTUM 'CONEFR'. IN THE GIVEN ORTHOGRAPHIC VIEWS, THE CONE IS DIMENSIONED USING GENERAL VALUES OF R, H, AND H1. FOR THE DEVELOPMENT, THE SLANT HEIGHTS VA AND VB ARE NEEDED AS WELL AS ANGLE, ANGL.

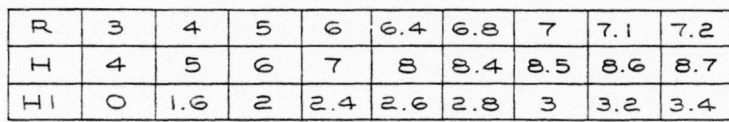

SLANT HT $VA = \sqrt{H^2 + R^2}$

USING SIMILAR TRIANGLE CONCEPTS:

SLANT HT VB: $\frac{VB}{VA} = \frac{H1}{H}$ OR $VB = \frac{H1 \times VA}{H}$

$ANGL = \frac{RADIUS}{SLANT\ HT}(360°)$ SEE PAGE 246 OF TEXTBOOK

SUGGESTED ARRAY OF VALUES

R	3	4	5	6	6.4	6.8	7	7.1	7.2
H	4	5	6	7	8	8.4	8.5	8.6	8.7
H1	0	1.6	2	2.4	2.6	2.8	3	3.2	3.4

VALUES IN CM

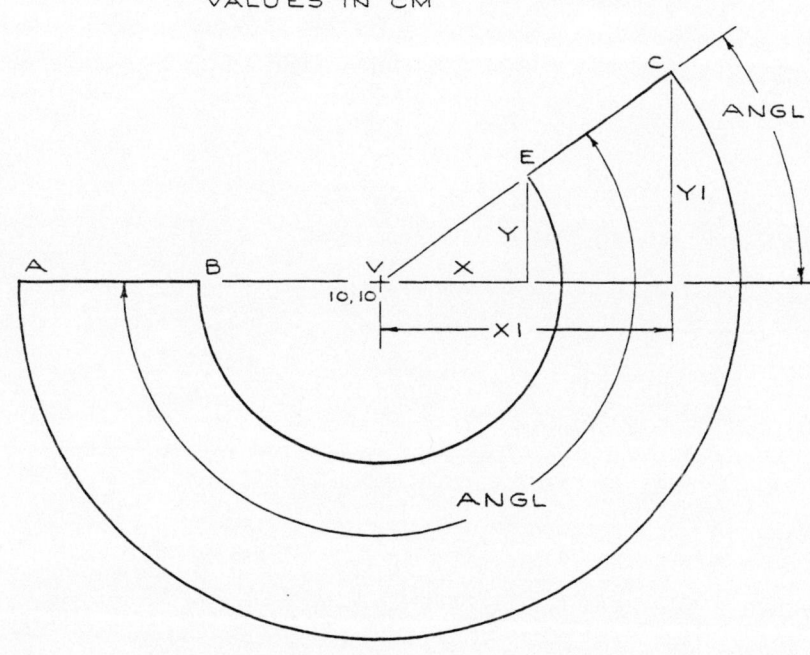

GENERAL COMMENTS

TYPE LOGIN
TYPE THE USER-ID, ME102

TYPE THE PASSWORD, ME102
AFTER OK, APPEARS ON THE TERMINAL, REQUEST THE LIST DIRECTORY; TYPE LD. SELECT PROJECT, TYPE CONEFR
VARIABLES OUT (1) THRU OUT (10) WILL BE USED TO STORE DATA OF UP TO 80 CHARACTERS EACH.
ANS *1 WILL BE USED TO STORE ONE-LETTER RESPONSES TO QUESTIONS; FOR EXAMPLE: Y FOR YES.
INTEGER *2 ID, I WILL BE EMPLOYED TO STORE WHOLE NUMBER VARIABLES. THESE VARIABLES ARE IN FLOATING POINT MODE TO ACCEPT DECIMAL VALUES. THIS ENTRY SETS LINE LOCATION AT 0. ID IS USED TO INDICATE WHERE TO LOCATE DATA OR ANS. FOR EXAMPLE: 0 FOR THE TERMINAL, 22 FOR PLOTTER. A SHORT EXPLANATION OF THE PROGRAM IS SHOWN USING PRINT STATEMENTS.

THE NUMBER 100 IS USED TO REFERENCE THAT LINE IN THE PROGRAM. THE STATEMENT 'CONTINUE' ALLOWS PROGRAM TO PROGRESS WITHOUT ADDITIONAL OPERATOR ACTION.

COMPUTER INPUT

```
LOGIN
USER-ID
PASSWORD
OK,
CONEFR

CHARACTER *80 OUT (10)

CHARACTER *1 ANS

INTEGER *2 ID, I

REAL *4 R, H, H1, ANGL, ANG1, RAD,
& RAD1, VA, VB, R1R

ID = 0

PRINT *, 'PROGRAM CONEFR'
PRINT *, 'USER PROVIDES SELECTED
& VALUES OF R, H, AND H1'
PRINT *, 'THE PROGRAM CALCULATES:'
PRINT *, 'DEVELOPMENT PLOT'

100 CONTINUE
```

COMPUTER GRAPHICS

IN FIGURE 1, A 45° LIGHT RAY HAS BEEN SELECTED TO
PRODUCE THE SHADOW OF THE ISOMETRIC PICTORIAL.
ANGLE (ANG) HAS BEEN CALCULATED AS 12° ON THE
PREVIOUS ISOMETRIC PROJECT (ISOSHA).

IN FIGURE 2, DIMENSIONS ARE GIVEN IN GENERAL
TERMS SO THAT A VARIETY OF SOLUTIONS CAN BE
OBTAINED.

FIRST THE PROGRAM IS ESTABLISHED TO PRODUCE
THE PICTORIAL. THEN CALCULATIONS ARE ADDED
TO OUTLINE THE VISIBLE SHADOW AREA.

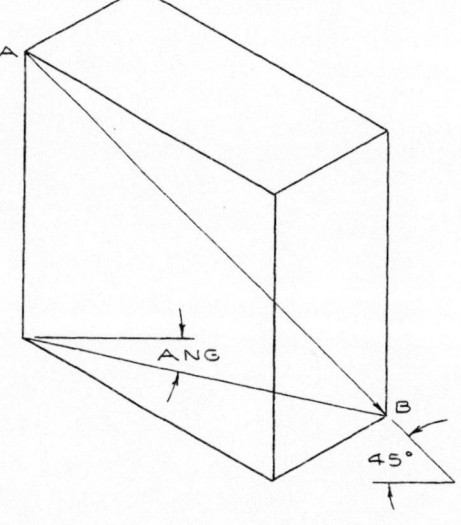

FIGURE 1

SUGGESTED VALUES

H	5	5.4	5.8	6.2	6.6	7	7.4	7.8	8.2	8.6
W	7	7.5	8	8.5	9	9.5	10	11	12	13
DP	3	3.4	3.8	4.2	4.6	5	5.4	5.8	6.2	6.6

VALUES IN CM

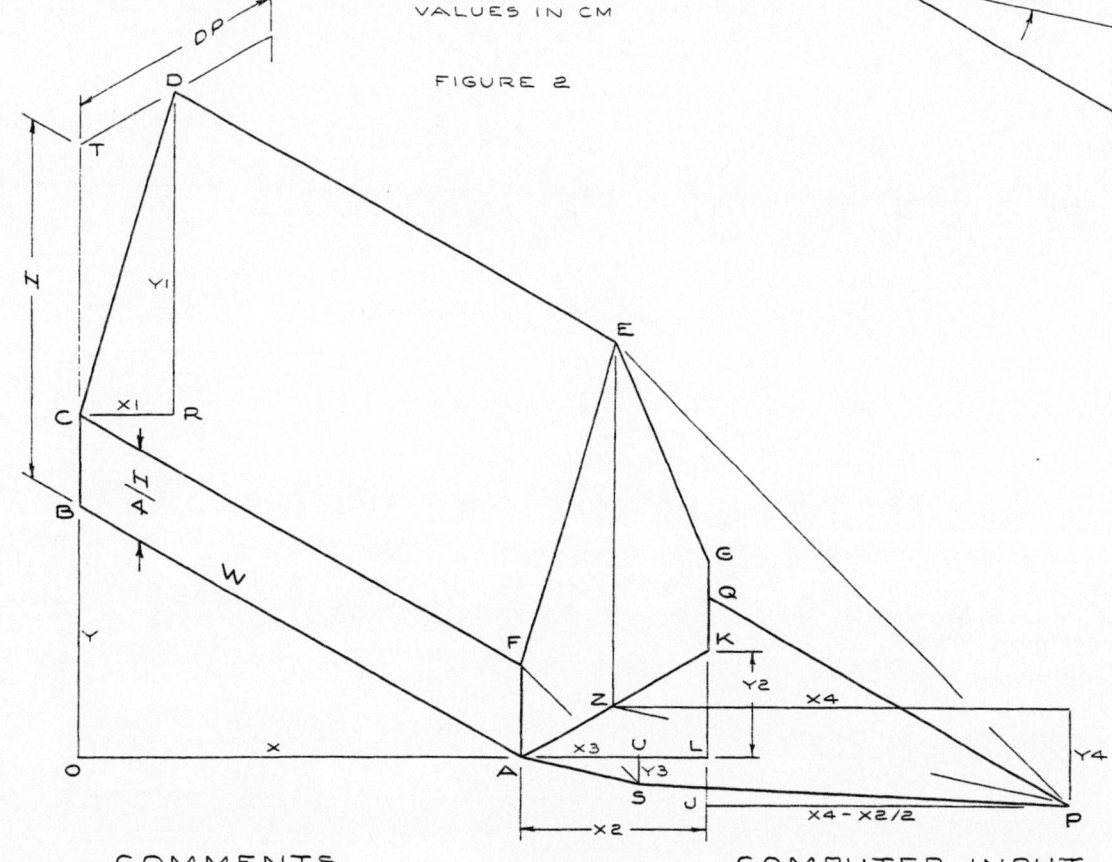

FIGURE 2

COMMENTS

TYPE LOGIN

USER-ID, TYPE ME102

PASSWORD, TYPE ME102

AFTER OK, SIGNAL APPEARS, TYPE
LD. THEN 102 MENU IS LISTED.
TYPE DESIRED PROJECT: ISSHAD.

PRINT SHORT EXPLANATION OF
THE PROGRAM.

USER IS ASKED TO SUPPLY H, W, DP.
FOR EXAMPLE: TYPE 5., 7., 3.
READ AND STORE THESE VALUES.
FOR ERROR, RETURN TO LINE 100.

COMPUTER INPUT

```
LOGIN
USER ID
PASSWORD
OK,
ISSHAD
CHARACTER *1 ANS
INTEGER *2 ID, I
REAL *4 H,W,DP,X,Y,X1,Y1,X2,Y2,X3,Y3,X4,Y4,
ID = 0
PRINT *, 'PROGRAM ISSHAD'
PRINT *, 'USER PROVIDES H, W, DP'
PRINT *, 'PROG. FINDS X,Y,X1,Y1,X2,Y2,X3,Y3,X4,Y4,QJ'
PRINT *, 'PLOT OF PICTORIAL AND SHADOW'
100 CONTINUE
PRINT *, 'USER PROVIDES H, W, DP
READ (I*, ERR = 100) H, W, DP
```

COMPUTER GRAPHICS (ISOMETRIC SHADOW)   25 g

THIS PROJECT ENTAILS A ONE-POINT PERSPECTIVE WITH THE RADIUS R AND DEPTH DP GIVEN. PT. A IS SELECTED AT X=8, Y=6. THE VANISHING PT. IS LOCATED AT X=18, Y=16. AN ARRAY OF THE TWO VARIABLES PERMIT MANY SOLUTIONS.

ARRAY

R	DP
3	2
3.4	2.4
3.8	2.8
4.2	3
4.6	3.6
5	4
5.4	4.4
5.8	4.8
6.2	5.2
6.6	5.6
7	6

VALUES IN CM

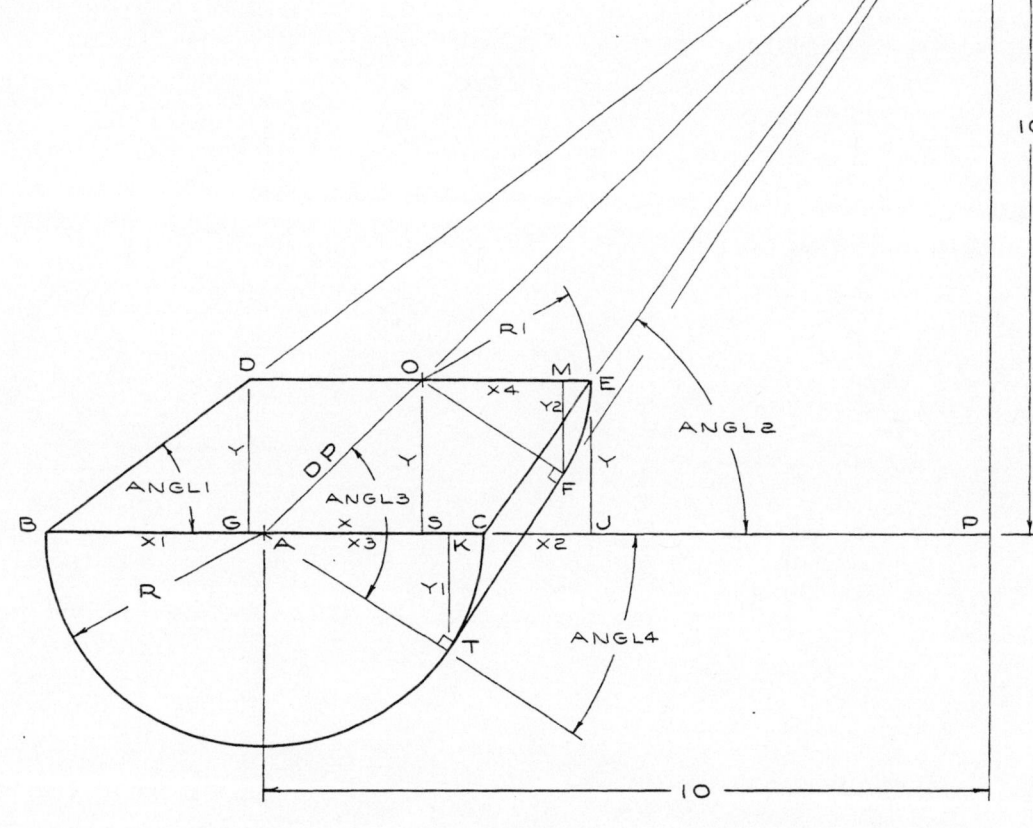

COMMENTS

TYPE LOGIN
USER-ID, TYPE ME102
PASSWORD, TYPE ME102
WHEN OK, APPEARS, TYPE LD
FROM THE LISTED MENU, SELECT AND TYPE THE PROGRAM, OPTPER

PRINT SHORT EXPLANATION OF THIS PROGRAM.

USER IS ASKED TO ENTER THE SELECTED VALUES OF R AND DP. FOR EXAMPLE: TYPE 3., 3.

READ AND STORE THESE VALUES. FOR ERROR, RETURN TO LINE 100.

IN TRI VAP, ANG VAP OR OAS = 45°
IN TRI OAS, X = COS(ANG OAS) × DP
AND Y = SIN(ANG OAS) × DP
IN TRI VBP, ANGL1 = TAN$^{-1}\left(\frac{VP}{AP+R}\right)$
IN TRI DBG, X1 = $\frac{Y}{TAN(ANGL1)}$

COMPUTER INPUT

```
LOGIN
USER ID
PASSWORD
OK,
OPTPER
        CHARACTER *1 ANS
        INTEGER *2 ID, I
        REAL *4 R, DP, R1, X, Y, X1, Y1, X2, Y2, X3, X4
        ID = 0
        PRINT *, 'PROGRAM OPTPER'
        PRINT *, 'USER PROVIDES R, DP'
        PRINT *, 'PROGRAM DISPLAYS'
        PRINT *, 'PLOT OF PICTORIAL'
100     CONTINUE
        PRINT *, 'USER PROVIDES R, DP'
        READ(1*, ERR=100) R, DP

        X = .707 * DP
        Y = .707 * DP
        ANGL1 = ATAN(10./(10.+R))
        X1 = Y/TAN(ANGL1)
```

COMPUTER GRAPHICS (PERSPECTIVE)   25 h